What We Take:
Memories of an Outdoor Biologist

By Dana Stewart Quinney

Hidden Shelf Publishing House
P.O. Box 4168, McCall, ID 83638
www.hiddenshelfpublishinghouse.com

Editor: Kerstin Stokes

Graphic design:

Interior layout: Kerstin Stokes

Library of Congress Cataloguing-in-Publication Data

Quinney, Dana Stewart
What We Take: Memories of an Outdoor Biologist

ISBN: xxxxx
ISBN-13: xxxxxx

Printed in the United States of America

Table of Contents

Table of Contents

Dedication

For Dr. Lyle Stanford,
who inspired me to pursue field biology.

Doc lecturing on the geology of Bruneau Canyon

Introduction

The foothills above my childhood home, the Fourhills.

I grew up when much of the world was open to children—the world of wildflowers, the world of animals, the world of rocks and weather and water. Now, it seems, for children, the world is closed. Only glimpses, carefully monitored, are available to them. Is that world safer for children? In the short term, yes. But in the long term, that narrow world is infinitely more dangerous, as restriction so often proves. And the wide world is infinitely more beautiful. This is my story, the story of a person who seems small in the great scheme of things, but who stepped into the real world as a child and never left it.

Growing up in Ketchum, Idaho, I became fascinated by wild things and drawn to learn about them. I became a field biologist, working all my life with wild plants and creatures in ecosystems both familiar and strange. From Australia's Great Barrier Reef to the high Arctic, to the Florida Keys, and the familiar sagebrush deserts of home–I traveled, and I learned.

To me, everything is a story. What monster moans at night in the rainforest? How do you attach lights to woodrats? What do spadefoot tadpoles eat? Will a mole take a shower?

Welcome to the sequel to Wildflower Girl. Come with me on my lifelong journey into little lives.

Shrew Tales, or Live Fast, Die Young

Part One: Little Lee

I was eighteen in 1962, and it was two months before I would enter the College of Idaho as a freshman.

Up Lake Creek, the hay had lain too long in the field, Dad said, but because it had rained a few hours after the hay was baled, the hay had to stay in the field until it dried out before being picked up and stacked in the shed, or else it would mold. So, the hay sat there for around ten days, drying.

The weather gods smiled, and finally it was time to pick it up. A friend of the family, Whiskey Chamberlain, and I managed to get the intact bales out of the field after a couple of days of hard work.

I drove Goliath, our 1936 ten-ton blue monster truck with thewooden bale ramp. Whiskey walked ahead and shoved the bales into the ramp, and every few bales, stacked them on the flat bed of the truck once they arrived there.

We took many loads to the shed and re-stacked them in the sweet-smelling dimness. On the second evening, we took a few minutes to rest and congratulate ourselves for a job well done— 1,600 bales picked up, stowed, and stacked. Whiskey clinked his bottle of beer to my bottle of grape Nehi as we sat on the last bale, a small celebration. We drove home in the long twilight, exhausted.

However, this year the twine used by the baling contractor seemed to be faulty, and many times it had snapped as Whiskey and I had tried to pick up the bales. We'd left a number of broken bales in the field. But it wouldn't do to let that hay go to waste.

I drove up Lake Creek the following morning to get those broken bales. Dad told me to bring them home in our pickup. Our horses could use that hay right away, and nothing would be wasted.

I packed a picnic lunch, dropped two bottles of Nehi into a little cooler, and took off.

Things went well, and I stopped for a righteous lunch. The sun was hot, but I was going great guns, as my gramps would say. After lunch, I drove to the last few broken bales.

As I picked up half a bale and turned to dump it into the pickup, Isaw something small and dark running away. I dropped the hay into the pickup. What was this little creature? A baby mouse?

Having lost his shelter, the tiny one circled back and ran to the only shade in the close-cut alfalfa—my shade. I saw him scurry over my shoe and felt him scrabble at my sock as if he wanted to burrow in next to my ankle. I reached down and caught him by the scruff of the neck, then lifted him for close examination. "Why, you're a shrew!" I said.

I had read about shrews. I knew they had to be fed often, every hour even, because of their rate of metabolism—but having raised many other small creatures successfully, I thought I could keep him as a pet.

I am Dana Lee, and he became Little Lee. I thought Little Lee very cute, but aside from his short-furred, cylindrical body, he was not appealing by most standards. His pinpoint eyes were so small as to be almost invisible. He had a ratlike tail, though short and furred, and his muzzle was distinctly pointed, with barely enough underjaw to let his mouth close. I am not certain which type of shrew he was, but thinking about it later, I've concluded that he was probably Merriam's shrew, one of the many species discovered by C. Hart Merriam in his pioneering travels of the West.

I was wearing a shirt with a button-down pocket, and I stuffed Little Lee inside and buttoned him in. With little effort, I caught a couple of fat grasshoppers and stuffed them into the pocket. I could feel Little Lee crunching the hoppers as I drove my load of broken bales home.

When I got home, I caught more grasshoppers for him, and heate them while I unloaded the hay and stacked it as well as broken bales can be stacked.

I grabbed a Mason jar from Mom's stash of canning jars in the basement and spent the last hour of daylight catching what I hoped were enough grasshoppers to last Little Lee until morning, and while he polished off yet more grasshoppers in my pocket, I made him two habitats: an at-home dirt palace in an old, cracked aquarium, and a travel coach fashioned from a shoebox with holes punched in the top for air and tied shut with baling twine.

I knew that Little Lee had to be offered live prey every hour or he would begin to shut down his metabolism and would start the process of dying. But I had been through intensive critter care several years before with the 18 hatchling toads I kept alive one summer, not to mention any number of baby birds and deer mice I had raised. To keep Little Lee going would be a challenge that I thought I could meet.

Every morning I dug worms for Little Lee. With my butterfly net, I caught moths, beetles, bugs, and (of course) many, many, MANY grasshoppers, and kept them in several jars to supply Little Lee's needs.

Because of his constant need for live food, Little Lee could not be left at home.

Little Lee went fishing with me. He went with me to my daily job of caring for the Sun Valley golf carts. His travel box sat on one of the carts each night as I swept them out and did minor repairs. He ate daddy longlegs and crickets from the corners of the golf shed, and dined on that morning's captured grasshoppers when I went home for dinner. At night, back in the dirt-filled aquarium, he ferreted out the earthworms I had placed into his home tank in the morning after he had been moved to his traveling coach for the day. He bathed in the dish of water there and would curl up to sleep in the pile of dry grass in one corner.

Weeks went by, midsummer weeks, late summer weeks.

Sometimes Little Lee rode in his shoebox coach, and at other times, he spent the day in my pocket. He went with me everywhere—to buy groceries, to mow the neighbors' lawn, for a horseback ride, for a hike, camping, even to a movie. As I sat in the darkened theatre, I could feel him in my jacket pocket as I moved a grasshopper from the left pocket to the right pocket, where he would grab each one instantly and crunch, crunch, crunch.

George and Annette Castle with my father, Clayton
Stewart (right) at one of their mining claims.
Yankee Fork of the Salmon River, Idaho.

Then one day my cousin Mark and I got long-coveted permission to drive Gramps' 1948 half-ton stake-bed pickup "over the hill" to spend the weekend with family friends George and Annette Castle, who in their little trailer were camped for the summer while George and his small crew logged up Thompson Creek, not far from the tiny town of Clayton.

We loaded our flyfishing gear and a food-filled cooler into the pickup. And I loaded a jar with both worms and grasshoppers. Little Lee rode in his traveling box behind the seat in the pickup. The pickup was slow on the steep summit, so it took us more than three hours to get to Thompson Creek, including the stops to give Little Lee some grasshoppers.

We arrived at the Castles' camp in early afternoon, and found that they had sited their travel trailer not far from a large and very appealing beaver pond. We gave Annette the ice and food that we had brought, and we broke out our fishing gear. I transferred Little Lee to my pocket with some worms and grasshoppers, put a mosquito fly on my leader, and set out to flyfish.

That day was perfect—warm and windless, the clear beaver ponds shimmering at the bottom of a very steep slope, the shade of fluffy green willows and leaning pines dappled on the meadow grass around the ponds, and the crisp air filled with that indefinable clean-water, grass-and-sedge, wet-willow smell that is my favorite scent in all the world.

Even though it was mid-afternoon on a warm day at summer's end, the fish were biting. I caught the biggest cutthroat of my life, a three-pounder, the fish thrashing the water like a silver windmill as I guided him in. I had no camera that day, but I can still picture him, the green back and rosy sides, the pale-ringed dark spots, the two bright orange cut-marks on his throat, the widely spread cinnamon tail with its wet black blotches. He was fat and prime and gorgeous. I let him go.

Annette and I cooked dinner, then after eating, all of us sat under the early stars talking quietly over our coffee as the darkness deepened.

Mark and I had found a grassy place in the nearby meadow and rolled out our sleeping bags for a time of starwatching before falling asleep. Mark and I were avid star watchers. I tipped a goodly supply of worms and hoppers into Little Lee's box and tied it shut for the night.

That night turned cold, down into the 40s. I woke at dawn and grabbed the jar of hoppers to feed Little Lee. But when I opened his traveling box, Little Lee was dead, lying motionless on his bed of soft grass. Live grasshoppers crawled aimlessly around in the box. He had eaten all the worms.

I knelt for a long moment there in the chilly meadow. What had happened to Little Lee? He had had plenty of live food. Had the night been too cold?

I knew that the average life of a wild shrew was about a year, and Little Lee had been an adult when I had found him in the hayfield. Had his small life simply run its course? With a sigh, I slipped on my shoes and began to look for a place to bury him.

I almost buried Little Lee. But then I thought, *I know shrews will eat carrion. What if another shrew comes along and is desperately hungry?*

So I searched for a little while and found a good place near the largest beaver pond. In the deep shade at the base of a willow, I laid Little Lee on a fine cushion of emerald moss and left him.

Part Two: A Nod to Thomas Rymer

A year later, summer came around again, and it was a difficult summer. I did well in my first year at college. My grades were fine. I loved college. But just before classes let out for the summer—well, let's just say that I didn't marry a boy who asked me and who didn't want to take no for an answer. I was astonished when he asked me. I had kissed him only a handful of times, nothing more. He was the classic nice guy, fun to be around, sweet, and considerate, but I knew it would never work. He kept writing and calling, writing and calling, and at last I felt I had to be unkind, and stopped answering. But I felt guilty, somehow. And distressed. Also, a long-term friendship with someone I thought I knew well had ended badly, a never-again-your-friend kind of ending.

Always a loner, I had a summer job that allowed me to be alone, and that was healing. And when I wasn't working, I took time to be in the mountains and hills, and that helped as well.

One beautiful July day I drove to the end of the road up the North Fork of Wood River, past the tumbledown Mormon camp and its graying, leaning cabins, and left my car a below the cirque and its ancient whitebark pines, where the pileated woodpeckers lived. I met no other vehicles, saw no one.

There were no established trails up to the cirque, but above me I could see its edge just below the knife-like ridge of the Boulder Mountains. On the other side was the upper end of the East Fork of the Salmon River country, a complex, long canyon system that would take four hours of hard driving to reach by road, but only a couple of hours to reach on foot, just on the other side of the gray granite ridge.

There was a patchy trail of sorts left by deer along a stream. The tiny stream was only a few inches deep, narrow enough to jump over, clear as air and cold as snow, tumbling in a little groove paved by pale granite rocks large and small.

I reached the grove of whitebark pine in the cirque and ate my lunch while being entertained by the woodpeckers. I didn't hike up the ridge to the divide, however. That day I didn't have the heart for it.

After an hour or so, I began to make my way back down the rocky slope.

Halfway down I found a small fragment of heaven.

The stream there fell in a shining sheet over a flat granite slab, a miniature waterfall a foot high, falling into a deep, rock-lined bowl the size of a kitchen sink. Gray granite ringed the falls and the bowl, the rocks separated by narrow ribbons of dark green and deep-red moss. Purple monkeyflower cascaded over the water, and a few inches from the edges of the stream grew scarlet Indian paintbrush, red-and-yellow columbine, pink heather, and deep blue penstemon. Bees and butterflies moved busily from flower to flower.

If ever there could be a place on Earth that had been enchanted by fairies, I thought, this was the place. I stopped in my tracks and lay full length in the moss and flowers, cradling my head on my arm. Surely it wouldn't hurt to rest for a little while. The sun was bright overhead, but the air was cool. Listening to the stream tinkling among the rocks and purling over the handspan falls, I fell asleep beside the water.

I woke suddenly, opening my eyes but not moving. The light had changed. I could tell that the afternoon had grown old. I felt reluctant to sit up and break the spell, and I felt strange. Falling asleep on an open slope in the middle of the day? For me, strange. Perhaps this place did have fairies, and I had been under enchantment.

And then I saw something like a gray leaf being carried over the handspan falls into the little pool. But it was no leaf.

Whatever it was, was swimming. Could it be a fairy? Did the alpine streams have dryads and naiads after all?

I lay motionless. A small, portly creature swam to the edge of the pool, hauled itself out, shook briefly, patted its round tummy with impossibly small paws, lifted its huge, pink hind feet, and scampered through the flowers to the top of the falls. What was this?

Before I had figured out what manner of creature it was, the fellow jumped into the water and rode the falls into the pool once more, and then swam again, a pointed head with tiny ears just out of the water, to the shore.

Then there were six of them, five more suddenly bobbing into view in the stream above the falls. They chittered and dove, glided over the falls, had a water fight in the middle of the pool, and swam toward the opposite shore, pinhead-sized pink noses turned up like snorkels.

The water was so clear that I could see the bottom of the pool where the creatures were swimming. One grabbed a pebble from the bottom, then dropped it. Another closed his rosy fist around a trailing grass head, and body-surfed in the strong current for a few seconds, then let go and did a rolling dive to the bottom and up again. As a group, they swam to shore.

While they were rubbing their wet faces with tiny front paws and smoothing the fur over their sides, my brain finally slotted into gear, and I knew. "Water shrews!" I said aloud. "You guys are water shrews."

I had, in fact, taken Introduction to Zoology at the College of Idaho during the past school year, and one of the books on our suggested reading list had been Konrad Lorenz's classic, *King Solomon's Ring*. And Lorenz had kept water shrews, European water shrews. I had no idea that water shrews existed in my own back yard.

I levered my head and shoulders up and lay there propped on my elbows, wondering why I had never seen water shrews. They didn't seem to notice.

I grew up in the Wood River Valley. I'd spent uncounted hours in and near its streams, rivers, and lakes, including the North Fork, and I had never even heard of water shrews. And yet, here they were in a miniature splashing mob, going over the falls again, ducking each other in the pool like mischievous children and doing bottoms-up dives just like my grade-school friends and I used to do in the pool at Bald Mountain Hot Springs when we were ten years old.

Furred a uniform dark gray above and white below, these little fellows were rascals, not fairies. On land, they were chubby little plodders, but in the water, they were expert swimmers like dolphins, small rodential dolphins. I knew from reading Lorenz's book that water shrews have a fringe of stiff hairs along the back of each hind foot that they can erect and use to help push themselves along in the water.

I watched the shrews for what seemed like hours. Splashes and dives, squeaks and furious paddling of the big pink feet, the group plodding like penguins along the bank so they could ride the falls yet again—I can see them now, a little peep show of unexpected happiness on an open hillside.

Suddenly a shadow swept across the slope, and I looked up: a goshawk. I watched the hawk glide down the canyon to where this small tributary met the main creek in a grove of Douglas fir and aspen. When I looked back at the stream, the shrews had vanished.

A chilly breeze lifted the bells of the columbine and shivered the heads of the Indian paintbrush. The day's last flash of sunlight glittered across the water. Then the sun blinked out behind the western ridge. I was alone, as they say, on the cold hill side.

I looked down the mountain and smiled. Somehow, the world had turned several shades brighter. In the long summer twilight, I could get to my car well before dark, and with luck I could drive home to the world of ordinary doings just in time for dinner.

The Tobacco Box

'm going to tell you a secret that I've kept for sixty years—well, part of a secret, not the whole thing.

Ubi sunt qui ante nos fuerunt. That's Latin, and it means "Where are they, those who were before us?" I took Latin in high school, and this phrase stuck to me like glue because I have always loved abandoned cabins, leaning fences, cars forgotten at the edge of a forest, purple-tinged bottles shining in the sunlight of a deserted sagebrush flat—and imagining who had lived in that cabin, who had built that fence, who had left that 1938 sedan in the aspen grove near that mine tunnel, who had guzzled the patent remedy in that bottle and had left it there to let the sun's rays, over many years, turn the clear glass to purple. What did those people do every day? Did they fall in love? Did they have a favorite horse? Sheepdog? Fishing hole? What did they wear? What did they do for fun? How did they make it through those long mountain winters? What were their dreams?

To me, the Pioneer Era seemed so romantic: people leaving homes, farms, and city jobs to head out into the wilderness to fashion new lives. How brave! And sometimes, how tragic. My mom felt an echo of this herself, and together we would often stop to explore those old cabins and forgotten corrals and decaying ranch houses. A purple bottle was a prize! Only bottles manufactured before 1900 turn purple with age. We found tiny opium bottles used by Chinese workers near some of the played-out mines, an 1880s dime near a fallen-in cabin up the Yankee Fork of the Salmon River, miners' test crucibles up Kinnickinnick Creek, bits of slag in the old rock kilns up Bayhorse Creek. We'd pause in the shade of some prospector's cabin, savoring and honoring the past. I'd see Mom put out a hand to the graying logs of a wall, stroking them. Our eyes would meet over the shadowy gulf of the past, and we'd smile.

When I was a freshman at the College of Idaho in 1963, the eco-era hadn't yet dawned. Few young people went hiking. For the most part, they didn't care to know the names of the flowers or the rocks, or where martens lived, or how to find flying squirrels. They did things together, town things, things that held little interest for me, and so during summers, it was rare for me to spend time with someone my own age.

I found a kindred spirit in our family friend, Annette Castle. Annette was a rockhound, eager to learn about fish, flowers, and animals as well. She was nearly my mother's age, and she and her husband, George, would spend the summers in the mountains, where he ran a small-time logging operation. For several years, each summer I'd spend a few weeks with them in some forgotten canyon at the edge of the true wilderness. George owned heavy equipment, and he'd fix up the old mining trails so, if you weren't too chicken, you could drive to that summer's logging camp. The camp would consist of four or five loggers living in tents or old cabins with their families, plus George and Annette, who in the summers lived in a 20-foot trailer. This trailer had long, deep grooves along the sides, where branches and rock formations had scored it as George, driving a Caterpillar bulldozer, had dragged it up some tiny road etched into the side of a mountain.

I'd help Annette cook for the loggers. We'd clear up after breakfast, make their lunches and our own, and then the rest of the day would be ours. We'd fish, hunt for agates and crystals, explore new gulches and canyons, and occasionally, we would drive into the nearest town for supplies.

Sometimes I didn't want to go into town, and sometimes Annette wanted to go to town by herself. This suited both of us. We'd leave the logging camp and Annette would drop me off at some side canyon where a tributary slipped into the Big River. I would fish or rock hunt, and get myself back to the drop-off point at a designated time late in the afternoon.

On this particularly bright July morning, I jumped into Annette's new but already mountain-scarred Cadillac, and we headed down to the main canyon of the Salmon River, where the only paved road led to the closest town with a supermarket, Challis, about thirty miles away. At my request, she stopped at a side canyon, very narrow and steep, with a thread of road heading up the mountain above a narrow, willow-choked creek. "Are you sure you want to fish here, Danny?" she asked. "With all these willows, it's going to be tough to cast without getting snagged."

"That's why I want to fish here," I said, hopping out with my fly rod. My creel was already heavy with a scrambled-egg sandwich and two bottles of grape Nehi. "I bet there's lots of fish. because of all these willows. I bet nobody else fishes here."

"Ok," she said. "I'm off for groceries. I'll pick up some more grape Nehi, too. See you later!" The Cadillac rolled back to the main road and disappeared around the bend. Below, the Salmon River was a sheet of silver in the late-morning sun. I turned toward the mouth of the dark canyon and fought my way through willows to the small tributary. I'll call it Fir Creek.

Fir Creek was clear as ice, and nearly as cold. Sliding over its rocky bed, Fir Creek was both narrow and shallow. At its widest, the creek spread five feet from bank to bank, and at its deepest? A five-gallon bucket would just fit into the deepest hole.

There were fish, cutties to be precise (cutthroat trout), in this creek. Fishing in a tiny creek with willows lacing together over the middle was the challenge I needed that day. Casting my fly into just the right spot without the backcast tangling in the willows was a serious test of skill, my kind of fun.

The first fish I caught astonished me. I'd heard Dad telling one of his friends about "purple cutthroats," cutties that were different from the ordinary (but still beautiful) rose-sided, dark-spotted cutthroats with the near-scarlet cut marks on the throat. The strange cutties Dad had described had spots and cut marks, but the cut marks on these fish, like the Fir Creek fish I had just caught, were golden, and the fish overall gleamed a bronzed-purple.

Fir Creek was full of hungry purple cutthroats. In two hours, I had caught my limit and hadn't even lost a fly.

While fishing, I had climbed upcanyon for perhaps a mile.

The south-facing slope above me showed a pale streak in its sagebrush and chokecherry bushes: the disused mining road. The north-facing slope appeared almost black with shadowy pines and firs.

The creek bed was becoming rockier and steeper. I was hungry, but the middle of a cold creek bristling with willows was not a good place to have lunch.

Mom and sister Vicki at an old cabin near
Clayton, Idaho, in 1960.

I climbed away from the creek and into the forest, looking for a nice fat log to sit on. I looked up; there was a level bench of ground above me. That would be a good place to have my lunch.

Through thick trees, I climbed to the bench—and dropped my fly rod on the forest floor.

A town. I was in a town.

Along the bench, and invaded by pine and fir, was spread a ghost town, the ghost of a town with what looked like the front of a store, a stable with tall double doors, and several smaller buildings. I took off my creel and laid it beside the fly rod, lunch forgotten.

I spent the rest of the afternoon exploring the town.

The first building I saw had indeed been a store, with countertops, shelves, and small barrels inside, with a stovepipe for a now-vanished stove. The building with the tall double doors had indeed been a stable, with several stalls, a rusted flat stove, and bins of shoe nails, plus one bucket of horse and mule shoes and another bucket of those strange half-shoes used for oxen. One wall was hung with harness and various bridles. The leather had been rodent-eaten, perhaps nibbled by generations of porcupines. One of the bridle-remnants had a strange bit—one half was standard, and the other half, a bar and side-ring, had obviously been a handcrafted repair, probably done right here in Fir Town, in this building.

On a countertop beside a pair of blacksmith's tongs sat a small metal box. A hole in the roof where a stovepipe had once fitted sent a round tube of sunlight into the stable, a beam of light that centered on the box, which glowed in yellows and oranges. It was a tobacco box, Pedro Cut Plug. Aha, I thought. The blacksmith chewed plug tobacco.

The outside walls of the stable held another surprise: skis! Perhaps twenty pairs of wooden skis had been fastened to that wall, long skis for tall adults, middle-sized skis, and short skis for children. Fir Town had been inhabited in the winters, too!

Eventually, I sat on the carpet of pine and fir needles at the center of the town and ate my lunch. Trees had grown thickly in the former road, both between the buildings and inside the buildings whose roofs had fallen in. In the tight canyon, the sun barely reached the roofs at midday, and by early afternoon, Fir Town slept wholly in shadow. I sat still for a long time, dreaming of people and horses in Fir Town, ox-drawn wagons, children playing marbles on the single street, people skiing down the canyon in the winter for supplies.

I glanced at my watch and realized that I had better get moving; we would have to put together dinner for the loggers, and I didn't want to keep Annette waiting. On impulse, I hurried back into the stable and took the bridle with the hand-repaired bit and the tobacco box. I scrambled down the side of the canyon and reached the bottom just in time to see Annette's green Cadillac pulling off the main road.

Annette pushed open the Cadillac's passenger door for me. "I got your grape Nehi," she said. The back seat was filled with grocery bags. "What are you carrying, Danny?"

I showed her the bridle and the tobacco box. "I—I found these up the creek," I said. Somehow, I couldn't bear to tell anyone, even Annette, about the ghost town. "And I limited out on trout. We can have them for breakfast."

"Great!" Annette said. She backed up the Cadillac, turned it around, and headed toward the highway along the big silver river.

Sixty years later, Mom and Annette, as well as George and Dad, have gone beyond the farthest ridge. I still have the old bit with its twisted remnant of hard leather. The Fir Town tobacco box sits on our dining room table. If you have read my SCAVENGER book series, you will recognize Fir Town; I wrote it right into that tale, though I changed its name and moved the location many miles.

The tobacco box.

And now you know part of an old secret. But where is Fir Town? That part I will never tell.

No Asylum

During my college days and ever afterward, I have loved field trips. When I enrolled at the College of Idaho, I quickly became part of the long-standing field biology program. Besides the long summer trips, shorter trips happened on a regular basis—several weekends and week-long spring breaks would find a few professors and a number of students in a car-caravan on the way to destinations like Death Valley, Bruneau Sand Dunes, Malheur National Wildlife Refuge, Ichthyosaur State Park, Sawtooth Valley, Uncompahgre National Forest, and many more places in the West. We would cook our own meals, tent-camp, listen to and give lectures—learning about plants, animals, history, and geology of the areas we visited. These trips were fabulous, unequaled in my experience, and gave me the foundation for my career in field biology.

For the cold and stormy spring break in 1965, our destination was Friday Harbor Research Station in the San Juan Islands, off the coast of Washington. I was excited by the chance to visit the famous marine research station. I was also excited by the opportunity to experience a week of learning and adventure with some of my friends and favorite professors. I remember that each of us students received one college credit for this week of intensive study, and that the cost of the trip was $25 per person, which included the station fees, college credit fee, food, and transportation. What a bargain!

Our first stop on the trip was a lunch stop in Pendleton, Oregon. We had just driven through the Blue Mountains, where the last of the winter's snow still lay piled three feet deep. We were glad to drive down the west face of the Blues and out of the snow. The sky was overcast, but there was little snow remaining on the Palouse.

Pendleton is an old sheep herding and farming town, famous to this day for the exquisite and almost indestructible wools from Pendleton Mills. As we dipped into the wide gulch where Pendleton has stuffed its houses and businesses, we drove through a residential neighborhood of little brick houses all in a row.

Suddenly the front door of a particularly tidy little house swung open so wide that the door smacked against the brick wall. As we stopped at a stoplight, I saw a man with no arms barrel out the door, dive down the front steps, and take off running in the yard. Why did the man have no arms?

Then I realized that the man was wearing a straitjacket, the sleeves wound tightly to his sides. Another man, shouting at the top of his lungs, ran out that same door after the first man. Mr. Straitjacket doubled his speed. In another couple of heartbeats, both had disappeared around the corner of the house.

"Holy moly, did you see that?" I asked my friends in the car. No one had seen the chase but me. I explained as the green light flashed on and we pulled away. "Do you think we should stop and call the police?" I asked.

"Well," said one student, "Pendleton is where the Oregon State Mental Hospital is. The guy was probably an inmate on a home visit or something. Don't worry about it."

"Yes," one of my girlfriends said, turning around to look at me from the front seat. "He can't do much harm in a straitjacket." From the look on her face, I could see that she didn't believe that I had seen a man in a straitjacket. We drove on.

On the far side of town, we found an informal roadside pullout with a garbage bin and a couple of picnic tables, so we stopped there for lunch. I was Commissary Director again that year as well, tasked with the responsibility of getting all the camping and cooking equipment together, keeping the trip budget, buying food, and supervising the cooking committees.

My lunch committee and I spread covers over the two tables. We set out bread, peanut butter, jam, butter, mayo, ketchup, mustard, lunchmeat, sliced cheese, and fruit, the standard on-the-road lunch of College of Idaho biology trips.

The temperature hovered around freezing, and we grabbed up sandwich components and ate our lunch while stomping around the rest area in an attempt to keep ourselves warm.

Nearby on the hillside loomed an institutional building, looking grave, stern, and antique. A student from Oregon pointed it out when he saw me staring at it.

"That's part of the Oregon State Hospital," he said.

"A hospital?" I echoed. "It looks like a prison."

"Well, it's one of the buildings of the hospital for the insane," he said, "so I suppose you could call it a prison. And look," he went on, "there are some inmates working on the grounds. See those men in the gray uniforms?"

I saw a few men, identically dressed in gray. Some were gathering fallen branches and piling them into a wheelbarrow. Others were emptying garbage cans. One was raking a space bordered by bricks that would probably be a flowerbed as soon as spring was farther along.

Two of the inmates edged closer to our picnic tables.

After talking to his fellow in low tones, one of them walked over to us and, leaning on his rake, said, "Hi! Where are you guys from?"

My friend Terry replied, "We're from the College of Idaho, and we're going to Puget Sound for spring break."

"Cool," the fellow said. "I'm one of the inmates here. They call me a trustee because they think they can trust me." He laughed. I felt a chill that had nothing to do with the icy wind. He pointed to a modern concrete building across the road. "See that building?" he asked us.

"Yes," I told him.

"Well, that's part of the hospital. That's where they keep us."

The building was massive, built from cement, dark-streaked where something had leaked and run down from the windowsills. This building was much newer than the one on the hill above us, but it had that austere, gray, institutional look, unmistakably.

"I have a roommate," the fellow went on, leaning on his rake and eyeing the food on our table. He nodded toward the man raking the flowerbed. "That's him. We're tight." He edged closer

24 - What We Take

to the picnic table.

Terry glanced at me. I nodded.

"Hey," Terry said, "would you like a sandwich? We have lots."

The man took off his thick gloves and shoved them under his arm. "Don't mind if I do," he said, and dropped the rake.

He stacked bologna and cheese an inch high on the bread and bit into his sandwich, chewing noisily. When the sandwich was half gone, he seemed to come back to the surface. He jogged Terry with an elbow and said, "You guys are all right. Can I have an orange?"

"Sure," I said.

He took two, shoving them deep into a pocket.

Suddenly he leaned so close to me that I could feel his breath on my cheek. "You guys are all right," he said again, "so I'm going to tell you a secret." I was glad that Terry was close by.

"You see that building?" he asked yet again. "That's where they keep us. I have a roommate and we have a room there. We don't like it there," he went on, suddenly seeming very focused. "We really don't like that place."

"I don't blame you," Terry said, stepping between me and the man. "It's not a very attractive building."

The man turned to face the big concrete structure, staring at the drab walls and rows of small windows. Traffic sped by on the road between us and the building, and on that dark roof, pigeons hopped and preened.

He began to speak very slowly, nearly chanting the words. "We don't like that place," he repeated. "We don't like that place." He grinned at me with yellow teeth. "And we're going to burn it down. Six of us. We're going to burn it down tonight. We have been planning this for a long time, and we're ready. You won't tell anyone, will you?"

"Of course not," Terry told him, edging toward the picnic table. "Well, so long. We've got to pack up now and get on the road," he said, but the man had already turned his back and was walking slowly up the hill toward the other inmates, rake slung over his shoulder.

"Jeez," Terry said. "He kind of gave me the creeps."

"Me, too," I said. "Come on, let's help put the lunch stuff away."

"Burn the place down," Terry repeated as he put caps on jars and slid a stack of unused paper plates back into a plastic bag. "That's a concrete building. How could they burn it down? Not possible, and besides, he's nuts. But the guy sure gave me the creeps." Terry gave a theatrical shudder and scooped up the knives covered in peanut-butter and spoons sticky with jam.

We caravanned through the Palouse, crossed the Columbia River, passed through sage hills, farmland, evergreen-covered hills, urban areas, and at last made it to our ferry on the near shore of Puget Sound. Before dark, we had landed at Friday Harbor Marine Station. We cooked dinner on the splendid stove in the cook cabin and sat down to a hot meal at long tables, the incident at Pendleton almost forgotten.

I loved our week at Friday Harbor: large purple starfish—with the occasional pink one—on the dark rocks at low tide, scarlet scallops with their snapping blue eyes and translucent filaments, little rock sculpins with spiny fins, frothy green-and-pink anemones in nearly every tidepool, barnacles raking the water with their tiny jointed arms, big wolf-eels with frightening teeth, a little orange octopus, weird worms with extendable jaws—we found all manner of creatures large and small.

We brought them back to the seawater-filled lab tables at the station and observed and classified them. Every evening after dinner, a student or two would give a talk about some aspect of the natural history of Friday Harbor (prepared beforehand). We had a few clear nights for taking star walks. Every evening, we'd sing and sing before heading off to the sleeping bags, so tired that we were almost staggering.

One evening, the cooking committee was going great guns on a spaghetti dinner and didn't need my help. A few of us wandered down to the dock, where we lay on the boards and looked down into the water in the deep twilight. Reflections from the station's few buildings rocked gently on the surface of the sea as the tide came in. So peaceful, I thought, staring down into the black water.

Something was down there, something that was not a reflection.

I blinked and shook my head to clear it. In the depths shone a pinpoint of light, light that was moving. Surely no one was diving in the dark. No bubbles broke the surface.

"Look!" I said to the others. "There's something down there, a blue light." Everyone peered over the edge of the dock into the black water.

"No," Terry said, "it's a red light, not blue. Wait—it's green."

Whatever the thing was, it was surfacing. The colors rippled through the rainbow—blue, green gold, orange, red, pulsating.

And there came another, and another.

The things, perhaps twenty of them, were now at the surface, rolling and flashing. "They're pickles!" someone exclaimed, and indeed, they did look like little rounded pickles bobbing at the surface of the sea in the near darkness.

"Wait," said Terry. "Look, they have two stringy things like combs—long things, longer than the bodies. And the colors are in bands up and down." He took a breath. "They're comb jellies!" he said. "Ctenophores!"

And they were.

We watched them for ten minutes as they flashed and rippled their neon colors while deep twilight gave way to darkness—odd, hollow relatives of jellyfish trailing their long combs as they manufactured light in their own tissues and sent it running up and down their bodies—red, blue, purple, green, orange, and gold.

Then a fresh breeze kicked up little wavelets, and in a matter of a few seconds, the comb jellies were gone.

We went into dinner without speaking. The sea is a wilderness and a mystery, and for ten minutes, we had been a part of Friday Harbor and its magic.

The remaining days at Friday Harbor passed quickly, and all too soon we found ourselves stuffed into vehicles, on the road home.

Near the end of our final day on the road, we passed through Pendleton, not intending to stop this time. Terry and I found ourselves in the back seat of a car together.

"Look," Terry said, pointing out the window, "there's the rest stop where we talked to that inmate and gave him a sandwich." He looked in the other direction, across the road, "And there's the building that he said . . ." Terry stopped in mid-sentence, staring.

The building was still standing, but now it had no windows and no doors. Heavy fingers of soot marred the concrete walls, reaching long black tongues from places where flames had burst the windowpanes and licked at the gray cement.

"I guess that guy and his friends really didn't like that building," I told Terry.

Terry nodded. "Must have been a pretty good plan," he said.

This story is dedicated to my friend Terry Uhlman, gone too soon. I will never forget you, dear Furry, Field Marshal extraordinaire.

The Burial Chamber

When she was twelve years old, my grandmother Lily and her little brother Douglas found a tomb, a Native American tomb.

They had been playing in a lava tube cave not far from their home in south-central Idaho, and had come around a sharp bend to a place where part of the cave's roof had fallen in.

They climbed over the rubble and squeezed into a place where they had never been before. Lily described it to me when I was a wide-eyed girl of twelve myself. Dim light. A stack of baskets. A pile of arrows. A bow. A bundle of large feathers. Some funny-looking sandals. A garment with dried and curled fringe. On a tray, a heap of seeds. Some rolled-up hides. The corpse itself, brown and shrunken, half-covered with a hide, the flesh of the face dry and glued to the bones, looking like leather itself.

Very excited, Lily and Douglas ran home to tell their father, Joe. They expected him to be as excited as they were. Lily imagined one of the lovely baskets sitting on the table in their kitchen, and Douglas wanted the bow and arrows for himself.

The next day, they led their father back to the rockfall and the cave. After Joe had taken a long look around the burial chamber, he said sternly, "We shouldn't be here, and you children can never come back. We can't take anything from this place."

"But the Indian is dead," Douglas protested.

"It doesn't matter," Joe Heller told his children. "This is a grave, and graves are sacred. All these things were put here to help this Indian in the other world. We don't believe what the Indians do, but I expect them to respect our beliefs and our graves, and so we should respect theirs. I forbid you to come here again."

But, of course, children will disobey. A few days later, Lily and Douglas went back to that tube cave, hoping to sneak into the burial chamber, but to their dismay, they found that their father had walled up the entrance with huge boulders—boulders too large and heavy for them to move.

Joe Heller made the children promise not to tell anyone else about the tomb, but when Joe had long gone to his own grave in the cemetery at Glenns Ferry, Lily told me.

***

Joe Heller

In the spring of my junior year at the College of Idaho, I thought it would be fun to look for that tomb. Lily herself had been back several times over the years to look for the tube cave, but had never found it. She gave me directions as best she could. Four of us students planned to make a weekend of it, driving my faithful 1960 Rambler American that I had named Coalie, my fiancé Eric, his younger brother Nick, and Nick's roommate, Gayle (a guy, not a gal). For us, May 1966 was all about discovery.

In one of my frequent phone calls home, I told Mom about our plans. Mom said, "I'm going to send you a package, Danny—something for your weekend." When the big box arrived on Friday morning and I carried it to my room from the dormitory's mailroom, I couldn't imagine what she had sent that could be so very heavy.

Mom was a genius. In the box were a cast-iron skillet, a huge package of family-caught trout packed in ice, and a pound of butter.

After our last classes, the four of us drove east into the lava-gulch country of the central Snake River Plain. We bounced up the narrow dirt road that led to the gulch that Grandmother Lily had identified, gathered sagebrush for a fire, laid our sleeping bags in a row, and began to search in a lumpy lava field for a tube cave where part of the roof had fallen in and the remainder of the cave had been sealed with a pile of boulders.

Twilight is long in late May. The sun wouldn't be setting until almost nine.

We found tube caves almost immediately. Many tube caves. Many, many tube caves.

Some were large enough to swallow all four of us walking abreast, and some were no larger than a mousehole, with all sizes between. Many had sections where the roofs had fallen. We felt that we were in the right place. We'd find that tomb, all right. What tribe had that leathery corpse called his or her own? Northern Paiute? Shoshoni? Blackfoot? We would take pictures, map the location, and tell our professors, who would know the right experts to contact. Treasure-hunting can be intoxicating.

The sun continued its journey to the west, and I began to think about building a cooking fire and putting on a pot of coffee.

Then Nick shouted, "Look! Here's a little tube just the right size for a person!" The three of us scrambled to him over the rocks. Nick had found an oval opening in the thick sheet of lava, a little away from the larger maze of caves that we had been exploring.

"Look inside," Nick said, stepping aside so we could peer down the hole. "Way down in there, there's some light. I'm going in."

We knew Nick all too well. As agile as a monkey, he was always climbing trees or rock outcrops and accomplishing strange acrobatic moves. Nick was slender and what some called "double-jointed." Sometimes I would swear that he could dislocate his shoulders so he could slip into slits between rock outcrops.

We tried to persuade Nick not to go into that narrow hole. I couldn't imagine, if he came to a dead end, how he could possibly back out. But Nick would not be dissuaded. Without so much as a flashlight, he wriggled into the hole, and in less time than it took to tell him to wait, the soles of his shoes disappeared into the blackness. He was gone.

A few minutes later and twenty yards from where Gayle, Eric, and I were sitting beside the hole, a hand flapped up from the lava. We ran to it.

"Hey," Nick said through a hole about six inches in diameter. "I tried going back, but that doesn't work, so I'm going ahead. See you later." We watched in dismay as his shirt moved past the hole, then his jeans, and finally, his shoes.

"We'll dig you out," we shouted to him. "We'll use the rock hammer. Stay there." But Nick was Nick, and Nick was gone.

The sun was setting. I built a pile of rocks at both small holes so I could find their locations after darkness came. "What are we going to do?" I asked the other two boys. I wanted to drive to the nearest town and get help.

"Have a little faith in Nick," both of them told me. "He'll be all right."

"If he isn't out of there by midnight," I said, deciding, "I'm going for help." The others agreed.

I carried dead sagebrush branches onto the lava and built my fire there. We didn't feel like eating, but we made coffee and sat sipping it in the long twilight. A full moon rose over the lava rim to the east. Coyote families howled from one rim of the gulch to the other, and our fire burned down into red coals.

Every few minutes, one of us would walk the few steps to the smaller hole, bend down, and shout, "Nick! Nick!" There was no answer.

I pulled on my sweatshirt and built up the fire.

Ten-thirty, and full darkness. As I sat on the hard lava, Coalie's keys seemed to be burning a hole in my pocket. Coalie herself was a gray shape some distance away at the edge of the lava; a sharp reflection of the moon shone in the top of her cab. The two boys lay on their sleeping bags, looking up at the stars.

The fire died down yet again, and I built it up once more. Gayle went to the hole and called for Nick. Then it was my turn. Then Eric's. Then Gayle's turn again. And I took my flashlight and gathered more dead sagebrush and built up the fire again and shouted Nick's name down the hole. And looked at my watch. Ten 'til twelve.

At midnight exactly, I stood, brushed off my jeans and said, "It's midnight. I'm going."

From the darkness to the northeast, I heard a faint shout.

"Nick?" I cried out.

"Nick?" the three of us shouted together.

We shone our flashlights northeast over the broken lava. That pale blob over there—was it moving?

The pale blob, almost two hundred yards away from us, was moving, and it was Nick extricating himself from a very small hole.

Stumbling in our haste, we ran to him.

"Hey," Nick said, sitting on the lava and dangling his feet into a hole that looked much too small for him to have climbed out of, "there was one place where there was such a tight kink in the passage that I had a hard time getting through, but it's really cool in there. Got anything to eat?" The flashlight beams showed a dusty, scratched Nick in a filthy, torn shirt. He grinned. "I'm starving."

By this time, the long-tended fire had burned down to a deep bed of coals, perfect for cooking. We ate trout fried in butter by the light of the moon and fell asleep soon after to the sound of coyotes singing above us on the rim.

Lily's little brother Douglas died in a train-hopping accident in 1914, when he was twenty. Lily left us in 1985 at the ripe old age of 91. Nick himself passed away almost 20 years ago.

Somewhere up a wide Idaho gulch, in an expanse of broken lava and half-collapsed tube caves, lies the burial chamber of an unknown tribesman or woman, still hidden after all these years. Joe Heller died in 1933, and the desert has kept his secret.

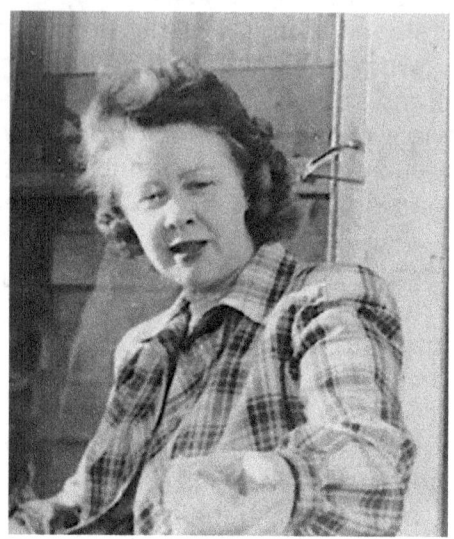

My grandma, Lily Heller Hicks

Revelation on Benjamin Hill

1965

I was off on my first long-term College of Idaho field expedition, late May through mid-August. The college had offered summer-long expeditions into Mexico for many years, and now, in the summer between my junior and senior years, was my chance.

There were twenty-seven of us, including two professors and a graduate assistant, and we traveled in a caravan of six vehicles pulling small trailers. One of them was a pickup pulling our commissary trailer. Each of us had a duffel bag of personal items, a sleeping bag, a small plant press, and a notebook in which to paste the pressed plant specimens. We were to study ecology and plant taxonomy, plus elements of animal behavior, geology, history, and culture as we encountered them. Our professors lectured every day, and we had both written and laboratory tests on the road. For this expedition, we would receive a full semester of credit hours.

On our journey we stopped nearly every night and set up camp—sometimes in parks or campgrounds, but more often simply in places where there was enough room to get us off the road safely. We did our own cooking, more successfully than you might think.

During the first ten days or so of the trip, we had camped our way south from Idaho, through Utah and small slices of Colorado and Nevada. We camped at Bruneau's Indian bathtub, somewhere in the pinyon-juniper hills of northern Utah, Arches National Park, Mesa Verde National Park, and in the pinyon country of northern Arizona. Our group had spent a day or two in Tucson Mountain Park, learning about the Sonoran Desert, where we would soon spend considerable time, and we had taken our first exam at Calabasas Forest Camp near the Arizona-Mexico border.

Photo by William H. Clark.

The day we crossed the border into Mexico was sunny and bright, as perfect a June day as one could wish, a carefree day, the day after a test. I was excited. Never before had I been outside my own country. We bought food in Nogales, and then headed south on the highway toward Hermosillo, traveling deep into Sonora.

I had always been fond of biology, particularly field biology, the attempt to understand wild things and wild ecosystems—but I was an English major. I loved reading and writing, had been successful at these pursuits, and looked forward to dedicating myself to many years of teaching students how to read for true understanding, and how to write with both grace and precision. I was going to be an English teacher, the best one ever, I figured. Biology was my minor. I had to have a minor, of course. But my life was mapped and planned out, and was going to be both sensible and glorious.

As the red Rambler American followed the pickup with the commissary trailer down the narrow highway, I glued myself to its window and took in the changing desert landscape. I got a head start at learning the plants and birds of the desert in Tucson at the Desert Museum, so I watched for them as we sped by. The large plants can be identified from a moving car window: spiky ocotillo, prickly pear cactus, tall saguaros, lacy creosote bush. Cactus wrens flew across the highway, and brown towhees and curve-billed thrashers, while caracaras soared above. Phainopeplas preened in the dry branches, flashing glossy black and dove gray. Cattle with their ribs showing nosed along the barrow pits, questing for something edible that wasn't dry and thorny.

The Sonoran Desert was a new world for this Idaho mountain girl. Even after all these years, this desert seems to me to be enchanted—not by fairies, of course—but perhaps by dried-up gnomes, odd and quirky, tossing out vast numbers of quirky oddities from their little workbenches.

As we drove south, the land grew more and more barren, the vegetation even more sparse, the homes and ranches fewer and farther between, a desert within a desert.

Little tin-and-adobe houses, scattered here and there in what seemed the least likely places, told me that this was a hardscrabble environment. Scruffy burros transported bags of something; lean chickens scratched in dooryards; vehicles were few.

In the earliest part of summer, the sun beat down with relentless energy. The Rambler, like most cars of that era, had no air conditioning, so we drove with all the windows down, panting, and I could only imagine living in one of those tin-topped houses in midsummer. I wondered where the people were getting water. I saw no streams or reservoirs; not even a well was visible. They must have water, I thought, and they must have a little to spare, because nearly every one of the tiny homes had a row of brightly painted pots or cans of flowers near the adobe walls—marigolds, bougainvillea, coral vine, and plants unknown to me, spilling color into the flat dirt courtyards.

Between one such house and the highway stood a very strange tree. Like nearly everything but the creosote bush, it had no leaves, and its smooth gray branches stood stark against the sky as the Rambler crested a gentle hill. The tree was small, perhaps fifteen feet tall. As we passed it, I almost cried out. I had never seen anything like that tree in my life before. The tree was in full bloom, nearly every branch and twig ending in a huge white flower. Sunlight glowed through the thin and delicate funnels of the round-mouthed blooms. *Morning glories*, I thought. *They are morning glories!*

Eventually we turned off on a little dirt road to the west, climbing a long hill. The sun was beginning to sink into the west, and a few clouds gave chase near the far horizon. We could see for miles. Not a building was in sight, not a car, not a burro, not a cow.

Photo by William H. Clark.

In the late afternoon we made camp in the desert scrub. A rusty barbed-wire fence paralleled the small dirt road on the south side, so we found a place to the north where we could get the vehicles off the dirt road without running into cacti. We slid the gas stove to one end of the commissary trailer, decided what we would make, and cooked dinner. Since the night was fair and mild, we didn't bother to set up our tents, but spread tarps on the ground for the sleeping bags.

As soon as my part of the setup work was finished, I sought out Dr. B, the botanist. As I described to him the mysterious tree, he began to smile. "Ah, yes," he said, leaning back against his pickup and slipping into lecture mode. "That's the morning-glory tree, *Ipomoea arborescens*. It's a true morning glory, family Convolvulaceae. You don't see many of those in the Sonoran Desert. That one may have been planted by someone on purpose. You'll see more of them, wild ones for certain, when we get down into the thorn forest. They bloom in the dry season, which is now, when they have no leaves. In the wet season, fall and winter, they have leaves but no flowers." He winked. "There's something about *Ipomoea arborescens* that spooks the locals wherever it grows. They call it witch tree, taboo tree, or *palo del muerte*."

Stick of death. Wow, I thought. *Stick of death. Muerte.* This had great appeal for my dark, Halloween-loving side.

Shortly after dinner, the other professor, Dr. J, waved his arms for attention. "This side is the boys' area," he said, pointing to the north beyond the camp. He pointed south. "And this is the girls' area." What he meant was "area for bathroom use," leaving east and west open for anyone to explore. He made this announcement at every camp, and without exception, we respected one other's privacy. The shovel, with an ever-present roll of toilet paper slipped over the handle, leaned against the commissary trailer, ready for use.

"Wouldn't you know it," I muttered to my friend, Cheryl. "The barbed-wire fence is on the girls' side."

"I'll hold the wire up for you, and you can hold it up for me," she replied. "It's going to be dark in a few minutes. Let's go." She took the shovel, and we negotiated the old barbed-wire fence on our way south to a place where we would be out of sight from the camp.

As the sun slipped behind a stack of narrow clouds near the horizon, we stepped carefully among the catclaw acacias, various cacti, and spiny wands of ocotillo, until we found ourselves behind a small earth mound, where we did our business quickly. This was not a place for a careless step, and extra care had to be taken during the actual bend-over, to make sure that no contact was made with any of those spines.

When we got back to the fence, Cheryl held the barbed-wire strand high so that I could slip under, and as I lowered my head, I heard a strange chittering. Louder and louder, the sound was coming from the east, and coming fast. I froze. This was a sound new to me.

Then above us, the sky filled with wings, black in the fading light—a flock of large birds moving unnervingly fast, birds as big as crows but bulkier, flying west. The unfamiliar cries came louder and louder, and then the birds were above us, just as a shaft of light from the setting sun broke free of the clouds.

They were *green birds*, sixty or seventy of them, each with a red crown and flashes of red on their shoulders. I had never before seen a green bird in the wild. The light went out as the sun sank at once behind the distant hills.

The fence forgotten, Cheryl and I stood and watched the birds until they became mere black dots disappearing into the dusk. *Parrots*. We had seen wild parrots.

Cheryl came out of the trance first. "Those are thick-billed parrots," she said. "Remember? We saw a pair in the aviary at the Desert Museum."

I nodded, too overwhelmed to speak. Here in this inhospitable, sparse, heat-ravaged scrub, lived parrots, a flock of parrots.

That night in my sleeping bag on the ground tarp, I could not sleep. All around me I could hear the quiet breathing of the others. Cheryl was fast asleep at my elbow, but I lay sleepless, staring at the roof of sparkling stars overhead, so many in the desert wilderness.

I had always pictured the tropics as lush and jungly, with great green leaves jammed as close together as possible— and in their deep shade, jaguars and tapirs and snakes on the ground, monkeys and parrots and exotic butterflies in the canopy, alligators, and piranhas and (of course!) tropical fish in the waters. Deserts were a gray blur beyond car windows, sliding past in uniform monotony, stingy with life and dull as dirt.

I had just discovered that deserts have their secrets.

The next morning dawned bright and hot. We began breaking camp at once. I rolled up my sleeping bag and helped with cooking our breakfast, while members of the other student committees were folding up the tarps and cleaning out the cars. Two fellows dug a deep hole and buried our garbage.

A cactus wren called from an ocotillo branch, sounding like

the harsh whir of an industrial sewing machine. White-winged doves cooed from the lacy shade of palos verdes, and turkey vultures soared on the first of the day's thermal currents. Tiny verdins hopped here and there, scolding. But there was no sign of the parrots.

"You guys got the dishwater ready?" one of the boys asked me. When I nodded, he said, "Good. What a godforsaken place this is. I'll be glad to get on the road today."

As I turned off the gas stove and shoved it back into its compartment in the commissary trailer, I felt indignant. How could Kevin, let's call him Kevin, feel that way about Benjamin Hill, Sonora? But, of course, Kevin had not seen all I had seen.

My little huff was short-lived, as Cheryl and I soon became involved in packing up the food boxes and securing them so nothing would rattle loose while the commissary trailer went bumping along the dirt road on our way back to the highway.

Within a few minutes, we were ready to depart, the cars and their small trailers now parked in a row on the dirt road behind Dr. B's pickup. Dr. B and the day's drivers huddled for a few minutes around a road map spread on the hood of his pickup, going over the day's route.

Nearly everyone else had already climbed into the cars, ready to temper the rising heat with some moving air.

I stood beside the red Rambler, taking in Benjamin Hill for the last time. Bird song was quieting as the heat of the day built all around us, and in the distance, shimmers in the air began to rise. "Not even as interesting here as those hills out of Tucson," came a girl's voice from the back seat of the Rambler. "I sure hope we have a decent place to camp tonight."

Ahead, the drivers' pow-wow was breaking up, and I could see our driver of the morning, Rudy, heading toward the Rambler owned by another student, Eric. I pulled open the front passenger-side door. And then I saw it.

Gliding smoothly on the bright red roof of the Rambler, exactly in the center, was a snail.

But this was not just any snail. The shell, twisted like a turret, was almost two inches long, white as old bones, glossy as porcelain. The animal inside, too, was white. The snail's flesh was translucent as opal. He left a moist trail on the red paint, and I wondered how he could possibly live here, where there was scarcely a leaf to be seen, where there was no surface water anywhere, and where almost every plant was spiny.

Rudy yanked open the driver's door on the opposite side of the Rambler and then he, too, saw the snail. Over the roof, our eyes met and held for a long moment of wonder. Without a word, Rudy, who was much taller, lifted the snail and placed him on the ground in the shade of a rock.

We got into the car without speaking. The Rambler was the last car in the caravan, and the others had already begun to move. Rudy started the engine and pushed in the clutch. I could hear the other passengers chattering in the back seat.

After we had pulled onto the highway, Rudy said, "I'm going to watch for snails from now on. I'm going to study them." I nodded. The snail had been eerie, a white, watery ghost, stranger than strange in this brown and thorny habitat.

Decades have passed since that camp on Benjamin Hill, and of course, much has changed since then. I looked up the thick-billed parrot online recently. Status of the species? Endangered throughout its range. The thick-billed parrot is confined to mountainous terrain, where it must have pine nuts to survive. Fewer than 2,000 pairs remain in existence, perhaps as few as 500 pairs. I read that thick-bills could not possibly survive in the desert, in places like Benjamin Hill.

I have tried in vain to find references to that snail with its long, tapering shell and glistening white body. To this day I can see the snail in my mind's eye, gliding silently, surrounded by paint red as blood, its mirrored reflection keeping the same slow pace, the trail of moisture behind it evaporating almost instantly in the sunlight.

After we had packed up camp, I considered the parrots and the snail, and suddenly found myself thinking about my plans to become an English teacher. I loved reading and writing, yes, and always had. And I had always loved the natural world as well.

But something happened then, in that hot car on that sweltering day, driving south into the heart of Mexico: the desert, the parrots, the translucent snail. *Passion* is what happened, stealing upon me as steadily as a snail traversing the roof of a red car, blooming in an unlikely place at an unlikely time, like the *palo del muerte*.

As the miles rolled by, I knew that I had to have both the words and the world. I would teach, yes. I would write, yes. But there was something more. I had to know about living things; I had to know their names, had to know their lives, had to know their world, had to work in their world. This was not a decision; it simply *was*. I now understand what changed: this is how passion does its work.

A small flock of thick-billed parrots flew across the highway, light glinting from their black beaks. I stole a glance at Rudy's sharp profile as he sat unmoving, his hands on the steering wheel clenched so tightly that his knuckles were white, and wondered what he was thinking.

Disaster Beach

Part One

Hot. The sun was very hot that afternoon, baking us in our un-air-conditioned cars to the tune of about one hundred and fifteen degrees—but that's not unusual in the tropics. We had been camping in pullouts along roadsides. But tonight would be different! We would camp on a beach.

On the 1965 College of Idaho field expedition to Mexico, I was already well on my way to earning credits equivalent to two lab science courses, concentrating on plant taxonomy and field identification of just about everything. Driving south, we were not far inland from the West Coast, deep into Mexico.

On this trip, Cheryl and I together were the Commissary Directors. For a considerable reduction of our tuition and fees, we had taken on the job of budgeting, planning menus, buying food, and organizing the daily cooking and cleanup. I learned rudimentary Spanish by going into little markets with a dictionary and uttering such profound statements as "*Quiero ses cebollas, por favor. Quantos? Gracias.*"

We had now been on the road for well over a month, familiar with packing and unpacking our things: sleeping bags, personal items, the library in the trunk of one of the cars, the girls' tent, the boys' tent, the professors' tents, and the big canopy tarp that covered the back half of our commissary trailer plus a large area of ground so that we could cook and eat in the shade or out of the rain.

The tents and canopy were not lovely Gore-Tex or light and airy nylon with rain repellent, and the canopy was not one of those totally waterproof blue tarps. All were made of heavy canvas—the kind of canvas that repels water only if you don't touch it when wet and break the air seal in the middle. And I do mean HEAVY canvas. The boys' and girls' tents were wall tents that could sleep as many as fifteen each, and the commissary canopy was huge. Each of the tents, folded and dry, weighed at least fifty pounds, and the canopy was even heavier. We had no light, flexible plastic tent poles. The poles were made of wood, with metal tips and joints.

The commissary trailer had been designed by students and a professor. I loved the old green and white monster. The boxy, tall fellow had two long, deep shelves on each side for storing canned goods, sugar, pancake mix, oatmeal, spices, eggs, Kool-Aid mix, pots and pans, and so on. The sides folded down into good-sized, easy-to-clean countertops, where a bunch of us could put together tacos, sandwiches, spaghetti, fruit salad, or work with whatever foods had been available in the local mercado that day.

In the center of the trailer was a long, hollow space with a pair of steel rails, and riding those rails was a four-burner gas stove, the cylinder of gas sitting in a special clamp behind it. To cook, one slid the stove to the back of the trailer, lit it, and voila—put on the coffee! When we were traveling, the two big tents and the canopy were piled on top of the rails behind the stove. When we needed the canopy, we'd attach the canvas to the top of the trailer, and ropes at each corner would be tied to trees (or if there were no trees, to our vehicles) to stretch it out, so we would have a dry, shady area for eating, studying, or listening to lectures.

The expedition rarely stayed in a camp longer than three days. Usually we moved every day, driving ever southward into Mexico, so we had become quite adept at dealing with these heavy tents, first packing our own sleeping bags and duffels into the small luggage trailers and then taking down the tents and canopy, a dozen or so of us working at that while others policed the camp area.

By this time, we felt that we were pretty efficient campers. We'd drive much of the day, stopping for lunch and also to look at birds, insects, flowers, trees, and to collect plant specimens to press for our classwork. We'd break for a lecture or two at places of interest. And in the late afternoon, we'd stop at places Dr. B had been using as one of the expedition camps for years. We'd haul out the tents and poles, raise the tents, stuff our sleeping bags and duffels into the tents, put up the canopy if needed, slide out the stove, lower the commissary trailer's sides, cook, eat, wash dishes, sing around a fire or a lantern, and fall into sleep sometimes even before we made it to the tents.

On this particularly steamy afternoon, we were driving through steep coastal hills, up and down, leaving the central high plateau, passing through coffee and banana plantations and the occasional tiny remnant patch of tropical forest.

We had been promised a beautiful camp on a beach for that night—a clean and lovely beach, with coconut palms swaying in the breeze, white sand, and a central lagoon. Our last camp had been in a none-too-clean space of weedy dirt between two sets of railroad tracks, so I was very much looking forward to the promised beach.

Fifty-five years later almost to the day, I cannot locate this beach on a satellite image, nor can I remember the name of the village close by, only that it was in the state of Nayarit, somewhere north of Puerto Vallarta. Perhaps the beach now exists only in memory—and I mean that literally.

The road was paved but single-lane, and we didn't dawdle on our way that day, because one of the cardinal rules of driving in Mexico is this: do not travel by night. Ever. Be off the road, be snug and unobtrusive, before night falls.

In 1965, Mexico was safe. The rural people were friendly and welcoming, and drugs had yet to rule the highways. What did rule the roads at night were horses, goats, cattle, sheep, and burros put out to graze on the roadsides, plus cars that may or may not have had headlights, and wooden wagons drawn by livestock, that may or may not have had a lantern hanging front or rear. We had to be camped by nightfall. Twilight is ephemeral in the tropics—night falls immediately after sundown, almost with a thud.

But—we made it to the beach while the sun was still well up in the sky. The people of the nearby village smiled and waved at us as we drove through the quiet central street. Many of them laughed as they pointed at our odd caravan of cars and trailers.

"Maybe they haven't seen too many cars from the outside," Cheryl, veteran of a previous C of I Mexico trip, remarked to me. "We must look odd to them, since they're laughing and pointing." A few days before, we'd driven a dirt track into a village called La Venta de Mochetilte (brisk wind) and had been told that we were the first cars ever to come there, so that was a possibility.

Cheryl is right, I thought. *That's why they are laughing and pointing.*

We turned toward the sea and found ourselves on a little sand bump perhaps ten feet above the beach, driving down into paradise.

The white beach unrolled below us like a landscape from a dream of heaven. This beach was a peninsula perhaps 200 yards long and 150 feet wide, a gleaming sand finger pointing westward into the blue Pacific. Graceful coconut palms ringed a large central lagoon. There was plenty of room for us to camp between the lagoon and the sea. I wondered if I could find some coconuts for dessert.

An unpaved track led down off the sand bump. We half-drove, half-slid down this little hill onto the beach. Here the part of the peninsula attaching to the mainland was a narrow neck, perhaps twenty yards wide. Looking from the back seat of the Rambler as we drove down through the sand, I noticed that a four-foot-wide chunk of the sandy road had caved into the sea, so recently that the sand was still slumping there, little threads of the white grains running down into the water. *Hmm,* I thought. A little prickle of unease ran down my spine.

Part Two

We parked our cars in a crescent between the sea and the lagoon, set up the tents, and began laying out our sleeping bags and duffels inside. We opened the commissary trailer and started dinner as the sun began to sink toward the sea.

Cheryl and I pitched in to help that night's cooking committee, as we often did, because we knew darkness was rapidly on its way. That night on the menu was fruit salad, with fresh watermelon, cantaloupe, pineapple, soursop, and mango. We fried up a big batch of fish, having bought them that morning in a *mercado* along the way.

One bit of camp rules from those C of I expeditions I later incorporated into my own expeditions for years and years: each person brought his or her own plate, cup, bowl, and utensils, and each person kept those items in a drawstring cloth bag. That way personal dishes could be washed and then hung on a bush or tree, inside the bags, to dry without being contaminated by dirt or flies. The bags themselves could be washed in a dishpan and kept clean.

Before long, dinner was almost ready, and I headed to the Rambler to get my dish bag.

On the way to the car, I stopped to take a few photos of the sunset. The sun was truly setting now, bleeding a path of fire into the sea.

And to the northwest, out to sea, billowed a towering thunderstorm. I could see lightning flashing and flashing. Distant thunder rolled gently over the waves. I took a few steps so I could frame the scene between the silhouettes of palm trees.

I ran to the car and grabbed my dish bag, but—my feet were wet. The Rambler was standing in four inches of water. I slammed the door, turned around, and nearly ran into Dr. B, who was also taking photos of the sunset.

"Dr. B," I said, just a little panicked, "the water is rising, and it's above the high tide mark. It looks like that storm is coming in."

He chuckled and said—for some reason I remember his exact words—"Don't be too concerned." He added, "Just get Eric to move his car." The Rambler was the car closest to the ocean.

I glanced over at the central lagoon. There, too, the water had risen right to the corner of the girls' tent. I told Eric and he ran to move the car to dry ground.

But the dry ground was disappearing, fast. The sun was a mere spark on the horizon. The light was fading. No longer distant, thunder slammed though the air above us. Weather happens fast in the tropics.

As if galvanized, everyone suddenly looked at the rising sea and understood. Frantically we ripped down the canopy and the tents, bundled them up complete with their contents, and shoved them into the center well of the commissary trailer.

We turned off the stove, unhooked its gas, and pushed it back into travel position. Somebody dropped the tub of fruit salad and what was left of the platter of fish into the trunk of a car. Dirty pots and pans were tossed onto the commissary trailer's shelves, and the trailers battened for the road.

Everyone came splashing to the cars, shielding their plates of food from flying sand. By this time, the wind was whipping the palm leaves, and it was nearly dark. There was no dry place left on the peninsula, none.

From somewhere to landward I heard a shout. "Hurry! Hurry! The road is falling into the sea!"

The brief twilight turned to indigo. Sand blew through the car windows and zinged across the windshield of the Rambler. Sea water began seeping in through the bottoms of the doors.

Eric gunned the Rambler and we started up the sand hump, the car wallowing, wheels madly churning in the sand. There was precious little left of the road up the sand hump to the mainland—the narrow neck of land there had mostly fallen away.

Before the last of the light faded, I looked out my window and saw a great chunk of the road crack, carve itself off, and fall into the sea just as the Rambler surged past it. Twenty more feet and we had made it.

We were the last car out.

Behind us, there was no road, no beach, no lagoon. Lightning showed me only palm trees standing in roiling water. Then the darkness took even that.

We drove slowly back through the village, seeing again the pointing and the laughter thanks to the light spilling out of the cantinas along the road. Yes, we were dumb, crazy Americans. But at least we had made it out.

Now, however, we faced another obstacle—driving at night as a caravan and trying to find a place to camp.

Part Three

I retrieved my dinner from the floor of the car between my feet and fed myself lumps of cooling fried fish and slippery fruit as we drove up and up through the little hills.

We passed a tall tree growing at the edge of the road, and I gasped. The tree was ablaze with fireflies—fireflies that made enormous lights. Every large tree was lit like Christmas Eve.

The sky was black velvet without a moon.

In the narrow beams of our headlamps, the gray road unrolled slowly as we crawled through the hills. Our headlamps showed us the eyeshine of burros, goats, and cattle—plus an unlit oxcart—and other moving things too small to identify. We met not a single car. And the great remnant trees leaned far over the narrow road, full of golden and winking stars.

After half an hour, the driver of our lead car spotted a little dirt road. Apprehensive, we stopped our vehicles on the highway while one of the boys got out and scouted the place. Five minutes later we saw him give a thumbs up, and slowly we turned off the road and found ourselves in a dirt clearing inside a patch of native forest surrounded by banana plants.

At the far end of the clearing, our headlights glanced off a little pool, and from the pool came a deafening cacophony of trills, whistles, groans, booms, and clicks.

I headed there with my flashlight and found hundreds of frogs of several kinds, green and gray and golden and brown—spotted, streaked, and plain, singing their hearts out, little throats puffed like soap bubbles.

The air was cool, a welcome relief from the heat of the day. It felt good.

We began to sort out our mashed-together gear. The storm hadn't followed us up into the hills, so we didn't need the canopy, but because of mosquitos, we set up both tents, ferreted out which of the jumbled sleeping bags and duffels belonged to whom, and began to settle for the night.

After all that excitement, Cheryl and I were glad to shake sand out of our clothes and crawl inside our bags. The other girls, too, got themselves ready for bed, and finally the last flashlight in our tent went dark just about the time the frogs stopped calling.

I was drifting off to sleep, listening to the quiet breathing all around me, when I heard a scream.

"Aaaaaa!"

One of the girls, still screaming, ran half-dressed out the tent door, at first forgetting to unzip it and nearly bringing down the entire tent.

"Something bit me, something bit me!" she was shrieking. I grabbed my flashlight and glasses; then Cheryl and I ran out the door into the dark. I was thankful that I still had on my shorts and t-shirt.

"Something bit me," is not a good thing to hear when you're in a dark tent in the tropics. That's a very scary thing to hear, especially when accompanied by multiple shrieks. My overtired mind went a little wild. A snake? Scorpion? Tarantula? Blood-sucking cone-nosed bug? In the half-darkness of the beach, we had thrown things into the trailers so wildly that almost any creature could be in there waiting to strike.

A silence stretched out, and I realized that everyone was looking at me. On the ground nearby I found a long stick. Armed with said stick and a flashlight, I went back into the tent.

Sleeping bags, pillows, clothing, and duffel bags from a dozen girls were tangled on the floor, the result of frightened girls pulling themselves out of sleeping bags and struggling to reach the tent door in the dark.

Working my way carefully inside, one by one I lifted items with the stick and moved them to a pile in a corner. This was going to take some time.

But I was wrong. After about five minutes, movement caught my eye. Something pale skittered across a pillow and hid at the end of a sleeping bag. My stick swiped the sleeping bag to one side, and at last I saw it.

There near the tent wall was a large, yellowish-pink, and very much embattled crab—a big guy, with a claw-to-claw span of about ten inches. He reared back on short legs and brandished his pincers at me. This crab wasn't going down without a fight.

Laughing, I plopped a pillow over the crab, reached under, and grabbed him by the back of the shell.

I showed the crab to the girls and then carried him over to the frog pond. The pool wasn't ideal for him, but I figured that maybe he could have a few tasty tadpole lunches before the lack of salt in the water got to him.

Tired girls reclaimed their tent, and a few curious boys went back to theirs. Our lights went out. "Holy gawk," Cheryl whispered to me in the dark, her comment on the day.

"Holy gawk," I whispered back.

More than fifty years later, that beach is gone, but Cheryl is still my best friend. Next time she comes over, I will have to ask her if she remembers Un-Paradise Beach, our narrow escape, the golden fireflies of the hills, and the stowaway crab.

The Lemon

I found it on Google Maps: a flattened space of dirt well off the west side of the *Carretera Internacionale*, the highway that winds through the steep hills south of Izucar de Matamoros, in Puebla, Mexico. In the late summer, our expedition camped there. A narrow escape brought us to those tropical hills.

A few days before, our caravan of six cars had been parked on a pullout near a highway a couple of hundred miles north of Izucar, all in a line. We were several feet off the road and had been travelling through rolling farmland when Dr. B had called for a plant-collecting stop. He had evidently seen a patch of something that we hadn't yet collected, a plant species we could learn from.

So, we stopped.

A bus, however, did not stop. Instead, it barreled down the highway and smashed into the last car in our caravan, Mrs. A's sedan. The impact flipped the small trailer onto the sedan itself, shattering glass and raining fragments onto the students still inside. The sedan butted into the back of the next trailer, crunching its own hood and fenders a bit.

I was in the red Rambler, two cars to the front of Mrs. A's sedan. I climbed out of the Rambler just in time to see the bus driver jump from the bus and run at full speed into a field, flinging away his uniform hat and jacket as he ran, disappearing into tall corn. We never saw him again. Had he fallen asleep while driving? We never found out.

A few of the students in Mrs. A's car had suffered small cuts from the flying glass, but no one had been seriously hurt. The trailer was a loss. The sedan was drivable, though the back windows were gone.

"Take all your stuff out of Mrs. A's car and trailer, and hurry! Get out your cameras and take lots of pictures!" Dr. B shouted then. "Get everything out of that car and stuffed into one of the other cars or trailers, right now! Take rolls and rolls of pictures!"

Photo by William H. Clark.

Puzzled, we did just that.

Police showed up minutes later. They interviewed us and the bus passengers, and then Mrs. and Dr. B were bundled into a police car and driven away. Mrs. A was fifty, much older than the rest of the students, and she looked terrified. I remembered that she didn't speak any Spanish. Dr. B did, however, so surely it would be all right—the accident hadn't been her fault.

We drove into the nearest town, and our graduate assistant, Ron, collected our rolls of film, located a photo-processing shop, and handed the proprietor a thick wad of cash. Ron told him there would be more if he hurried. I hadn't been much concerned until I saw Ron's face. His jaw was set, his brows drawn together. When I asked him what he thought would happen, Ron didn't even look at me. "Come on," he said. "Follow me, everybody. We'll find a place to camp out of town."

Mrs. A and Dr. B were not at our camp in the dusty woods that night. We ate dinner and went to bed early, sobered from the events of the day. The following morning, we drove back into town and parked on the street near the photo shop.

On that street, we sat in our cars for hours. Finally, the proprietor of the photo shop came running out. He handed Ron a bag, and Ron handed him another wad of cash. "You guys stay here," he said to us, and his voice was grim. "I'm going to the courthouse."

"Why are you so worried?" I asked as he got into his vehicle.

"Don't you know about the Napoleonic Code of Law?" he said. "In our court system, you are presumed innocent until proven guilty. Here, you are presumed guilty and must prove yourself innocent. And that's a big bus company we'll be fighting." He drove away, disappearing around a corner.

The rest of us sat in our vehicles and cooked in the sun for two more hours. Then, just as we were beginning to wonder what we were going to do for the night, we saw Ron's car coming around the corner. He pulled up; then Mrs. A and Dr. B climbed out.

"Are you OK?" we all asked as we crowded around them. Mrs. A looked shaky, as if at any moment she would fall over, and we guided her back to Ron's car, where she sat in the passenger seat with the door open.

"I'm OK," she said, "but my car is gone. I'll never get it back."

"Tell all!" we demanded.

Dr. B gathered our attention. "I thought we were jailbirds for sure," he said with a smile. "The bus company claimed that we had stopped in the middle of the highway, all of us, and that it would have been impossible for the bus to have avoided hitting Mrs. A's car. But then Ron showed up with your photos, and the photos won our case. It was so obvious that we were WAY off the road. Thank God," he added under his breath.

"What about the bus driver?" I asked. "Did he testify?"

"He wasn't there," Dr. B said. "The bus company claimed that he was too upset to testify."

"He ran!" I said.

"Yep," said Dr. B with a wry grin. "And he didn't even take any photos."

"But what about Mrs. A's car?" one of the girls asked.

"Confiscated," Dr. B said. "Yes, it was fixable and drivable, but it's gone. We'll never see it again. They don't need much of an excuse to confiscate things, it seems. That's why I had you get your belongings out before the police arrived. But this isn't quite over. We won't be allowed to leave Mexico until we get another car, and to do that, we have to leave Mexico."

We didn't understand. Dr. B had led many Mexico expeditions, and with the voice of long experience, explained to us that Mexico was extremely protective of its automobile industry, and had strict laws against people bringing cars in from the USA and selling them in Mexico. Therefore, anyone who came into Mexico with a car, had to drive out of Mexico with a car. Otherwise, officials would catch that driver at the border as they tried to leave the country, and they wouldn't be able to. The attempt would mean significant jail time. "The insurance will pay a little, but not much," Dr. B added. Indeed, as we later found out, it paid only a few hundred dollars.

"This can't be!" We argued while around our fire at camp that evening. "It doesn't make sense! We can squish into five cars and three trailers for the rest of the trip."

"Doesn't matter if it doesn't make sense," Dr. B said. "Here's what's going to happen. Mrs. A is going to take a bus back to the border. She has a court document allowing her to leave Mexico and come back into Mexico with a vehicle. Ron will go with her."

The next morning, we dropped Ron and Mrs. A at a bus station. Mrs. A still looked terrified; she kept very close to Ron and seemed very small as she followed him through the glass doors into the dark building.

The rest of us drove south. "We're going to make camp south of Izucar de Matamoros," Dr. B told us. "We'll have to stay there at least a week, to give Ron and Mrs. A time to get to Nuevo Laredo, cross the border into Laredo, Texas, buy a car, and drive all the way back. It's 800 miles each way."

We bought food and filled up with water in the town of Izucar. As we drove the winding highway up into the steep hills, we caught a breathtaking view of one of the world's most beautiful volcanoes, Popocatepetl, 18,000 feet of snow-topped majesty. It felt very odd to see the snow-capped peak above the giant cardon cacti and other tropical vegetation of the nearby hills.

Dr. B had chosen our campsite where we'd wait very well.

We saw wild begonias and Leclancher's buntings, found exotic scarab beetles, talked to friendly inhabitants, and could collect plenty of wood for our nightly campfires.

Eventually, a flattish green Edsel showed up, and out hopped Ron and Mrs. A. The thing was huge and low, a dinosaur.

C of I Mexico Expedition camp south of Izucar de Matamoros, Puebla, Mexico. Photo by William H. Clark.

"It was what I could get," said Mrs. A testily. She told us that the Mexican border authorities had grabbed Ron as he tried to cross over and had locked him up in the local jail. Ron had driven his own car into Mexico and was thus "trying to leave without it." Mrs. A had to make her way across the border alone. In Laredo, she found a used-car dealership. She had never bought a car by herself before, but somehow, she had managed to buy this Edsel. Ron talked his way out of jail, she picked him up in the Edsel, and they headed south.

"You did your best, Mrs. A," Ron said soothingly. "But the thing is a lemon. We've stopped to fix something on it in almost every town we passed through on the way here," he told us, shaking his head. "Dr. B, I don't think we should continue further south, the way we planned. I think we should start heading for home before the thing falls apart completely."

Dr. B nodded. "I think you're right, Ron," was all he said.

The next morning, we packed up our southernmost camp and headed north toward the border. We stopped to fix something on that Edsel, all the way back to Idaho.

The Place of Fear

I t was very, very hot that day in east-central Mexico. Mrs. A's Edsel was making it, though with frequent stops for minor repairs. Midsummer had come, and with it, 110-degree temperatures.

Even through the roofs of our cars, the noon sun beat down like a hammer upon me and my fellow expeditioners as we drove to *El Tajin de Veracruz*, the archaeological site.

In 1965, back-country Veracruz was a landscape of small hills peopled with nodding oil derricks and white-clad native Mexicans going about their rural business. By this time, we had been away from home for two months.

On that sweltering midsummer day, our caravan of cars arrived at the Pyramid of the Niches via a tiny and rutted dirt road, that was somehow both muddy and dusty at the same time. We passed scores of conical hills overgrown with small trees and shrubs woven together with tangles of tough vines. Passionflower, coral vine, lantana, and other exotic flowers laced themselves through the wild growth, taking advantage of the fact that the great rainforest trees had all been logged away, allowing the lesser plants riotous access to the sunlight.

We had been told that we were going to spend the afternoon at El Tajin, an ancient holy site of the Maya. The chief feature of El Tajin was the Pyramid of the Niches, with its 365 square recesses, each just large enough to hold a person's head, a cluster of candles, or a small statue. We could not wait to see it.

We bumped around the final bend in the road, and there stood the Pyramid of the Niches, gleaming like a carved bone in a setting of emerald. Rearing its stair-stepped profile sixty feet into the humid air, the pyramid had been cleared of most vegetation, and its cracked gray limestone stood in high contrast to the shimmering green that stretched to the horizon on all sides.

For several minutes we drove among hills that were not hills. They, too, were pyramids and other man-made structures, their forms covered and softened by a thick blanket of soil and vegetation. This was a place of many, many ancient buildings. In 1965, only the Pyramid of the Niches and a ball court had been excavated.

The Pyramid of the Niches (right), El Tajin, Veracruz, Mexico.
Photo by William H. Clark.

We scrambled from our cars into the intense heat of midday, midsummer Veracruz, and began to set up our folding tables. Not another soul was in sight while a flock of parrots flew over. Insects hummed and heatwaves trembled in the air as we set out the lunch things.

We worked slowly, as do those who are accustomed to moving about in tropical heat. We made sandwiches and ate our lunch, carefully stowing everything afterward in the green-and-white commissary trailer.

After lunch, we gathered around one of our professors for a short lecture on El Tajin. Nothing was known of the civilization that had built the city, only that El Tajin was said to mean "place of the twelve thunder gods," or "the place where the thunders meet"—or simply "lightning."

Near the foot of the Pyramid of the Niches was a deep, round hole perhaps twenty-five inches in diameter. Into this, a tall, heavy wooden pole had been stepped. Dr. B told us that local descendants of the Maya, with colorful wings strapped to their arms, would don harnesses and climb this pole, attach a line to a ring in the top, and while suspended, would "fly," circling the pole, to the ground. At that time, no one knew why the flight of the bird men had to be here, and what this festival meant to both these ancient people and their modern descendants.

The professor's short lecture ended, and it was time for us to explore.

Most of the students climbed the pyramid. I walked all the way around it, marveling. What had the 365 niches once held? Votive offerings of food and drink? Small statues? Oil lamps or candles? Sacrifices of some sort? Were the niches used one at a time, or all at once?

A little later, I found myself staring at a flat-topped hill not far from the Pyramid of the Niches. The small hill was shaped like a building, but at one corner near the bottom of the hill, there was a high bump, or sharp mound, like the buried prow of a ship, covered with soil and vines like nearly everything else in the landscape. Behind the mound, the towering clouds of a tropical thunderstorm rolled, building slowly for the usual afternoon downpour.

The entire hill had a threatening aspect, and I turned away with a shiver. Something about that sharp bump made me fidgety and uncomfortable. Uncomfortable enough to remember the feeling that chilled me to this day. I moved away.

Close by in another direction I saw a series of low, rectangular mounds, with chunks of shaped white limestone protruding from the vegetation like broken teeth from green gums. I wondered what marvels were hidden under their thick green blankets, and how long it would take me to excavate one of these buried buildings if I started now using the shovel that the expedition carried. I moved into the shade of a small tree, feeling that I was melting into my shoes.

"Come on," Cheryl called. "Look over here. Here's the ball

court Dr. B told us about."

Cheryl led me to a sunken area with a long, rather narrow field bounded by low walls of the same pale gray limestone as the pyramid. Behind the wall was an up-sloped area filled with soil and ranked with slabs of stone: seating.

Cheryl and I walked onto the overgrown playing field and talked about the football and baseball games that were a familiar part of our lives at home. El Tajin's people would have brought food and drink (especially drink, since it was so hot much of the time), and perhaps would have made bets on the games, shouting, and screaming for their players.

Or perhaps not.

"Look," I said to Cheryl. "The walls are carved!

We began to examine the carved reliefs on the low stone walls. The carvings were complex and sinuously curved, with grooves wandering here and there to fill and decorate every inch of the space on the surfaces of the stones.

Between intricately chiseled borders we made out arcane images: a parrot-headed god, other figures too strange to be even partly human, human heads being eaten by great fish, temples with many doors, men holding writhing snakes, and men with snakes for tongues.

We stepped back a little, and more images fell into place. We saw a procession of men in patterned cloaks and buckled belts, wearing large earrings and impressive headdresses crowned with tall feathers. Some of the carved fellows were more elaborately dressed than others and were larger.

These are the important guys, I thought. *These littler guys are less-important, or maybe they are the ball players.*

"Over here," Cheryl said, beckoning. "Here are some long things, like staffs, carved into the stone, all lined up. Maybe they tell the date or the season or something."

Cheryl and I walked slowly along the walls for some time, the heat beating down on us, our very small shadows falling as tiny pools of shade around our feet. We examined the carved walls of the ball court, and suddenly both of us lifted our heads to look into the distance.

Beyond this court were the outlines of more ball courts, and still more, and more yet, covered in shrubs and vines, their outlines appearing as bumpy lines and hollows under their living green shrouds.

The top of my head burned and drops of perspiration dripped from the tip of my nose and splatted onto my shoes. It was too hot to think. I had to sit down.

I stumbled across the court to the far wall and pulled myself up onto it, centering myself in a tiny blotch of shade from a small tree. As I lifted my head, I could see the carved procession of ball game officials on the opposite wall, wavering in the steaming sunshine as if they were alive.

There, in a moment of immense clarity, I saw what I did not see when I was close: one of the smaller men had his arms bound behind his back. He was being offered to a tall fellow in elaborate dress who had his arm extended toward the prisoner's throat, his hand holding a pointed knife. Human sacrifice.

These ball games were nothing like our American football and baseball. These games had been part of a rite, a ritual.

The artist who had carved that scene had told me this: *human sacrifice happened here, and the victim was unwilling.* As clearly as if he had screamed it to me across the ball court, the long-dead artist had spoken directly to me, a young woman from another time and place, and I had understood.

Lightning blanked the sky and clouds swept across the sun, stirring a breath of wind. Thunderheads piled above us, black-bellied and huge.

Cheryl and I walked back to the cars and began to call for the others. We knew we had to get back to the highway before the afternoon rain made the little road impassable. I, for one, was more than ready to leave.

***

The El Tajin we saw in 1965 as a pyramid, a single ball court, and green mounds, is now an excavated city with many temples and buildings, with more than a dozen ball courts, covering more than two square miles of the green-mounded jungle we saw then. El Tajin has a paved access road, maintained walkways, a parking lot, a visitors' center, and guards. People are everywhere. It is very much not the same. It is estimated by historians that the city dominated its region during much of the time between 600 and 900 A.D. But who lived at El Tajin? Who built it, who decorated it, who spilled blood for sacrifice on its pale gray stones? In 1965, it was thought that El Tajin was a place of the Maya, but now historians say that El Tajin is not Mayan. The people of this city are lost; we know nothing about them except what we can learn from their stones.

The ball-court artist told me they were a people very familiar with fear.

In an internet search, I found photographs of the flat-topped green mound that had so disturbed me in 1965, the mound near the Pyramid of the Niches. Now excavated, the mound I remember is a low, rectangular building of gray limestone. The strange, unnerving bump I had seen near this mound has also been excavated.

The bump, uncloaked, is a massive pillar carved from black basalt. The pillar is a statue of the god of death, whose head is a strong, open-mouthed skull.

Christmas in July

Now summer was old, and the 1965 Mexico Expedition continued to head north, only a few days south of the Texas border.

It was the 25th of July. On this date, it was the tradition for C of I Mexico expeditions to celebrate Summer Christmas.

We were to make a special meal to celebrate, and one of the professors had found a piñata at the *mercado* in Saltillo, the last town we had passed through on our way to that night's campsite. He bought bags of wrapped candies to fill it.

This piñata, attired in rows of pastel crepe paper, was shaped like a burro. The professors had filled the burro with candy. The plan was to have an after-dinner piñata-bash, with blindfolded students taking turns trying to smash the piñata using a long, heavy stick.

"*El burro is muy perezoso,*" commented one of the students, this being one of the few phrases of Spanish he had learned before we left Idaho. "The burro is very lazy." He said this on many occasions unrelated to burros or laziness.

We camped in a glorious place beside the Rio Salto in the Mexican state of San Luis Potosi, just below the waterfall.

The Rio Salto has limestone as its bed, and so the water is colored the most beautiful turquoise blue, clear but for this tint. At El Salto, or the falls (the jump), the river then dropped about a hundred feet into a magnificent, rounded pool, and as the river flowed on, the water went into and out of several huge cuplike limestone basins, each one forming a little waterfall of its own from five to fifteen feet high. Arching from the banks still are giant Montezuma bald cypress trees, elegant in ferny foliage, draped with festoons of gray Spanish moss. (The Rio Salto falls has since been captured in many pipes for generating hydropower and no longer thunders over that limestone rim.)

We camped about twenty yards from the steep bank of the river, and one hundred yards from the falls itself. Beyond the river, thorn-forest hills filled the eye in every direction.

Rio Salto, San Luis Potosi, Mexico. Photo by William H. Clark.

We set up the tents and commissary trailer and cooked our special dinner. We moved quickly, for darkness gives little warning in the tropics and the sun would set soon.

At last, someone tied a rock to a rope, tossed it over a tree branch, and hoisted the burro piñata off the ground. We gathered in a circle below it.

Blindfolded, I was handed the long stick and took a swipe. Everyone laughed. I'm five-foot three, and even with a long stick, I couldn't reach the piñata. I pulled off the blindfold and tied it over Cheryl's eyes. She couldn't hit the piñata, either.

The guys were taller. One of them hit the piñata a glancing blow, and it swung above us like a pendulum. Finally, one of the boys JUMPED as he swung the stick and connected a solid *thump!*

The burro pinata slipped from the rope and tumbled—all the way down to the river. Someone climbed down the clifflike riverbank and retrieved it, sodden, now with flopping ears and colors running from the crepe paper. We found some candies on the ground in a trail leading to the riverbank, but most of the piñata's candy had floated away on the current and (I hoped) fed the fishes downstream.

During the piñata affair, we hadn't noticed the clouds moving in, and suddenly we realized that the sky was darkening above us. The time of day was sunset, but the early darkness told us we were in for a storm.

Quickly we lit our lanterns and retired to the tents. It would be sticky and dank packing up camp in the morning, but for the present we were tired and full of warm food. We wouldn't miss one night of singing around the fire (or lantern, as we would have done here; in many of our forest camps, every bit of fallen wood had already been harvested). It would be fine to turn in early.

We had already been through many a tropical rainstorm in our heavy olive-drab canvas wall tents. The canvas was thick and the wooden poles strong. If we were careful not to touch tents' roofs and sides, the air seal would not break even in the hardest rain, and we'd stay dry.

That day, we had pitched the tents on a slight slope, where the ground began to rise away from the river. Inside the girls' tent, the slope seemed rather pronounced. But we figured that if we weren't too active during the night, we wouldn't slide toward the door.

I set the Coleman lantern in the middle of the tent floor while we girls unrolled our sleeping bags.

I fluffed up my tiny flannel-covered pillow and deposited my glasses in one of my shoes, as always. Gratefully, I settled into my bag, leaving the side unzipped.

The first patterings of raindrops soon sounded on the roof of the tent, followed by a great flash of lightning and crack of thunder, so close, it drowned out the voice of the falls and the constant hiss of the flowing river.

Then the deluge fell, pounding into the roof of the tent and pouring off the sides. We weren't concerned. We always trenched around the tents while in the tropics, and we had also trenched this time, digging a drainage channel from one corner down to the riverbank

I called, "Ready?" and turned off the lantern.

The thunder and lightning died away, but the rain continued.

I woke intermittently during the night. The rain—if you could call it rain when it came in sheets rather than drops—pounded on.

As dawn began to bleed in through the heavy tent fabric, the rain tinkled to a stop. I was warm and comfortable, and thought sleepily, "This is so cozy. My feet feel like they are floating."

Suddenly, something nudged my shoulder. I felt for it, and my hand closed over the stiff wire handle of the Coleman lantern. The lantern had fallen over and lay on its side, floating.

Floating.

Abruptly, I sat up. Where I clutched that wire handle, my hand was wet. I flailed in the dimness and found my shoes. They, too, were floating. I snatched my glasses and put them on, then found my personal flashlight and clicked it.

During the night, the Rio Salto had risen over fifteen vertical feet and 20 horizontal yards and had entered our tent. And even in the few seconds that this took to register in my sleep-fogged brain, I could see that the water was rising.

"Wake up! Wake up!" I shouted.

The six or eight feet of tent floor on the door side was almost a foot deep in water. Only a small portion of the uphill part of the floor was dry.

"Get everything out!" one of the girls shouted.

"No!" shouted another, splashing. "There's no time. Get out. We'll pull up the stakes and drag the whole tent."

I slipped on my shoes and stood in the chaos for a moment, with splashing all around me. I was calf-deep in water when at last I dashed for the door. All of us girls made it out.

The falls were not a murmur, but a roar.

Somehow, we tugged the tent stakes free of the mud and then, working together, pulled the entire tent back and back and back, away from the water.

From somewhere behind me, I heard an engine start, and another. Headlights gave us spidery shadows as we dragged the tent. The boys were moving the cars and trucks away from the river.

We worked without speaking. At any rate, the unending crash of the falls made it impossible to hear voices. We pulled that tent uphill a hundred feet, a hundred and fifty.

A watery sun broke through the treetops and showed us the prospect of a new day.

The gentle, calm Rio Salto with its limpid turquoise waters, was gone. In its place brawled a brown monster. Uprooted trees, giant cypresses, whirled downstream. I watched as the thick trunk of one broke the rim of a limestone basin and crashed down into the next. The falls looked as if every drop of water in the world were foaming over that cliff to bury us. It was terrifying.

Soberly, we packed up our wet things as best we could and left the Rio Salto. Soaked and shivering, I climbed into my assigned car and asked the girl in the front seat to turn on the heat. "This would never have happened at home," she said. "Never. How were we supposed to know this would happen?"

Lucky, I thought to myself. *We were lucky*. And then I knew, knew something I carried with me for many years, to many expedition camps along many tropical streams.

"It's the tropics. Anything can happen."

Panya and the Sky

In 1967, my husband Eric was a graduate student in entomology at Oregon State University. We were living in a small apartment in Corvallis, on a hill at the edge of town.

The entomology graduate students were a close-knit group. Many had come from other countries, fully committed to learning agricultural entomology. Some of these students were on scholarship from the governments of their own countries; in those countries, agricultural entomologists were sorely needed to deal with crop-eating pests and insect-transmitted crop diseases.

One of these students, Panya, hailed from Thailand. This was his first time in the USA, and he acclimated rapidly. Panya was bright, articulate, kind, funny, and soon became a friend. His apartment was in the same complex as ours, so we saw him often. I remember his ringing laugh when he presented my husband with a silk tie. "A tie from TIE-land," Panya said, bowing.

One spring weekend, we ended up doing a favor for a different friend. Another graduate student in the group, Jim, needed some help. He was studying parasitic insects that lived in bird nests—swallow nests, in particular. For his thesis, he was collecting pinches of litter from swallow nests and Berlese-ing them.

Need to get small insects from soil, forest leaf-litter, or bird nests? Place the material into a metal funnel and place the funnel over a jar. You may leave the jar empty or half-fill it with alcohol. Place the funnel under a bright incandescent light bulb, and you've got what entomologists call a Berlese funnel. There's no need to sift the material, no need to pore over it under a dissecting scope, no need to crush or disturb the little guys within while you are trying to extract them. Typically, small creatures that live in litter will move away from heat and light. In a Berlese funnel, they wriggle or crawl farther and farther away from the light and heat (down the funnel) until they fall into the jar. Bingo—no fuss, no muss, no detritus—and after a few hours, there are the little critters for you to study, perky and undamaged, or (if you use alcohol), perfectly preserved.

Jim had to get to barn swallow nests. He had nearly killed himself trying. Of course, barn swallows nest on barns. They nest on barns and cliffs, anywhere with considerable height above the ground. Climbing sheer walls is dangerous.

However, Jim had finally found a barn swallow colony that was easily reachable. A few miles south of town, a low wooden bridge crossed a sluggish stream that meandered through a small marsh.

"If you wade into the stream, you can go under the bridge, and the swallow nests on the girders underneath are reachable," he told us one night as we met in one of the entomology labs at the university. "There are dozens of nests. You can reach right into them."

"Cool," I said. "But you're taller than either of us, Jim. Why can't you do the collecting yourself?"

A flush of dull red crept up Jim's neck. "I itch," he said, looking away. "I got into the water, waded under the bridge, and collected a few pinches of stuff. But then my legs started to burn and itch, so I got out of there. By the time I got home, I had these horrible hives all over my legs. They lasted for days. I couldn't even sleep." He looked me in the eye. "Please," he said, holding out a stack of brown-paper envelopes, "I only need material from 20 more nests."

We took the envelopes.

"I'm all set up here," Jim said, indicating a long row of Berlese funnels on a countertop. "It'll be easy. It's already dark. You go at night so you don't have the birds dive-bombing you. They have nestlings, and if you disturb the birds during the day, in the fuss some nestlings will fall out of the nests and drown. So, tonight will be perfect. There isn't even any moonlight. Reach into the nests, get a pinch of the stuff inside, and put a sample from each nest into a separate envelope. Just scoot the babies aside. The adults will bail, but that's OK. They come right back."

"All right, Jim," I told him. "But if we get hives, I'll never forgive you."

In the dark, we drove to the marsh south of town and located the low bridge. Holding flashlights, we eased ourselves waist-deep into the cold water and headed under.

Our lights showed us brown blobs bobbing along on the surface of the stream. "That's nutria manure," Eric said. "Maybe Jim is allergic to nutria manure. I hope we aren't."

Without incident, we located the nests.

It was startling for the swallows, but lovely to look into their sharp-pointed little faces, worm our fingers under small nestlings fuzzily warm in their soft gray coats, and pull out small clumps of material to put into the envelopes.

We busied ourselves under the bridge for some time. At last, we had filled every envelope—a good thing, because the chill of the water was getting to us. Time to go. We waded out from under the bridge.

The world had changed.

The sky was orange. Great veils and curtains of brilliant orange filled the midnight sky with light and reflected on the water throughout the marsh, as far as we could see. The curtains faded and re-formed, flowing and changing. Starlight burned sharply through the translucent veils.

Astonished, we stood motionless in the cold water for an unmarked space of time.

Eventually, we hauled ourselves up the bank, changed into dry jeans there in the car, and drove home.

It was one in the morning.

From the parking lot, I could hear the phone ringing in our apartment, ringing and ringing. As we reached the door and fumbled with the key, the ringing stopped. Then the ringing started up again, just as we came through the door. Who would be calling us in the middle of the night? Had someone in the family had an accident? Worried, I lifted the receiver.

On the other end of the line was Panya. "Oh, I'm so glad. Finally, I have reached you," he said. "I don't know what to do.

Do you know where there's a shelter? Can I come with you?"

"What's wrong?" I asked him. Panya sounded panicked. I didn't understand. Panya had always seemed so calm and logical.

"What's wrong?" His voice rose, trembling. "What's wrong? It's nuclear war! At last, it has come. Please, can I come with you? I don't know what to do!"

"Sure," I said. "Come on over."

Almost at once, Panya was at our door. We sat him at the kitchen table and explained to him about aurora borealis, the northern lights.

Intelligent, well-educated Panya had never heard of the northern lights. I suppose that many people who live in the tropics don't know about them. Schoolchildren there probably learn about monsoons and tsunamis instead. At any rate, aurora phenomena had been left out of Panya's schooling. Being a reasonable fellow, he calmed down right away.

The three of us took cold drinks outside. Sitting on a concrete barrier at one end of the parking lot, we watched the northern lights bloom and fade, bloom and fade until the creeping gray of dawn banished them from the sky.

Bear Canyon Rat

In the summer of 1967, we moved to Tucson, Arizona, where my husband Eric began work on his doctorate in biology at the University of Arizona. We loved living in the Sonoran Desert, and I still have a great fondness for those prickly hills and shadowed canyons.

One day we went for a walk in Bear Canyon in the Catalina Mountains. It was a bright, hot afternoon in early summer. We had planned a leisurely, interesting hike, and what we got instead was a lesson.

Bear Creek had stopped flowing several weeks before. For weeks every summer, the stream exists only as a series of disconnected pools in the tumbled rock of the canyon. Here and there the creek sports wide, sandy beaches that will again be underwater when the late-summer monsoon rains begin, and the creek flows once more.

On that day the small, isolated pools shimmered in the intense heat, threaded with the lacy shade of mesquite and palo verde trees.

We had a long hike up the canyon and eventually retraced our path down the trail beside the rocky streambed, more than ready for a drink from the cooler in the car.

At one point we came to a small beach of golden sand at the edge of a small, oval pool.

The pool was perhaps ten feet long and four feet wide. The clear water showed the sandy bottom about twelve inches down. Looping toward us on the surface of the pool was a swimming snake, a medium-sized gopher snake, so we stopped to watch him. He made it to the nearby bank and curled there to rest behind a rock.

I became aware of a blink of movement on the far side of the pool. There sat a large rat grooming her fur, which was damp and spiky, as if she had gone swimming a little while before. Beside her was a younger rat, perhaps one-third grown. This pair would be mother and child. The young rat, too, had spiky, damp fur.

They were plain, ordinary brown rats, their ancestors having sailed from Europe on a great adventure, jumping ship at some wharf in the Americas long ago. These two were far from any human habitation and were living wild. Their bright black eyes, clever pink hands, and sharp, adaptable brains could well make a living almost anywhere in the world. Bear Canyon was a good place to live for a rat—cacti and their fruits; many insects, lizards, and worms; grasses and their seeds; and abundant places to nest in rocks, trees, and cacti.

With a final pass of her paws through her whiskers, the female launched herself into the pool. At once, the baby followed. He was not as graceful but managed to keep up. They had nearly reached the opposite side of the pool when the dragon awoke.

The gopher snake raised his head from the sand. He saw them; mother and child were swimming straight for his rock. Before I could react, the snake swam out to meet them.

Mother Rat was wily. At once she turned to put herself between the snake and her baby, but the young rat was less adept. He bumped into her side, then bounced away and was exposed.

The snake thrust out his head and grabbed the baby rat by the ribcage. Baby rat squealed and thrashed, but the snake's bite held, and the serpent turned back toward the shore with his prize.

But it wasn't over. Mother Rat swam furiously and caught up with the snake before he reached the sand. She bit him on the back of the head.

Letting the baby go, the snake turned on Mother Rat. He bit her three or four times in quick succession. She floundered, sinking, and I thought she was done.

The snake picked up the baby and swam once more for the shore. But as the tip of the snake's tail slid past Mother Rat, she bit and held on.

This time the snake kept swimming, no longer willing to let go of his prey. Like someone climbing a rope, Mother Rat clutched and bit her way up to the center of the snake's body, and finally, he let go of Baby Rat. The small one, stunned, began to sink to the bottom.

Mother Rat dived beneath the little one and lifted him to the surface even as she was being bitten by the snake; the snake had turned back to attack her. She shoved the baby toward shore and fought for her life.

Eventually, Baby Rat floated to the beach, crawled feebly from the water, and collapsed.

Mother Rat and the snake flailed and thrashed in the water. She was bleeding from both front paws, there was a tear in her skin over the ribcage, her ears were lacerated, and her tail was deeply gashed. At one point, the snake held her head under water for so long that I thought she could not possibly survive, but she tore away and at the last moment snatched a breath of air.

She was bitten many times and could have fled, but she doggedly refused to slip away each time the snake turned from her; he had in mind easier prey. However, Mother Rat was now tiring, and after a time, the snake eluded her and swam toward the baby on the beach.

The snake was intent on Baby Rat, not the mother. He swam past Mother Rat, picked up the baby from the sand, and began to swallow.

By the time Mother Rat swam to shore, only the pink hind feet of her baby were visible. I thought she would give up in order to save herself. She was seriously injured, and surely the baby was beyond rescue.

Mother Rat flung herself upon the snake and bit him through the eye.

With his mouth full of Baby Rat, the snake could not bite her. Relentlessly she bit and bit his head, until he lay limp on the sand.

Immediately, Mother Rat turned to her child and pulled him from the snake's mouth.

Baby Rat was slimy and limp. He did not seem to be breathing. Mother Rat collapsed near him on the sand and began to lick him. Then she pummeled him with her front paws.

Suddenly the little one gasped. He shivered and turned to look his mother in the face. With both front paws, she held his head and groomed his whiskers over and over.

Shouldering him up, she half-pushed, half dragged Baby Rat to shelter, some openings in the knotted tree roots near the edge of the water.

Bending down, I could see the two of them huddled in the half-light of their small grotto. Mother Rat patted her child all over and licked him. Then, seemingly satisfied, she began to lick her own wounds.

Stepping lightly over the dead snake, we left them. Soon they would return to feed.

Pete:
A Life in Black and White

Pete was a magpie, and he came into our lives because he fell from a nest.

In 1968, Eric's brother Nick was teaching at Bishop Kelly High School in Boise, Idaho, when one of his students brought in the ugly, nearly naked baby bird. Nick, being Nick, immediately fixed a cage in the classroom for the baby magpie and began feeding him. The name his students chose, for whatever reason, was Pete.

Pete thrived on the good food and the attention he received in Nick's classroom at Bishop Kelly. But as it happens, school let out in the spring and Nick had places to go, things to do. So, we agreed to take Pete in as our own. Nick brought Pete from Idaho, where magpies were common, to our home in Tucson, Arizona, where there were no wild magpies.

In the carport of our rental house, we built an aviary from chicken wire and 2x4 boards. Six inches of clean sand from a nearby arroyo filled the bottom, and inside we anchored a big juniper skeleton from up on the mountain. We added a large water bowl, wide enough for Pete to wade and bathe in. Tucson's mild winters would be comfortable for Pete outdoors, and the shaded carport would help him endure the hot summers.

Pete moved in and took over the aviary. That summer he became round and glossy, and grew the long, black, iridescent tail feathers characteristic of his kind. He would eat almost anything, but his favorite treat was raw liver, and he would do backflips when he saw me coming through the sliding glass doors from the house with a little plate of liver strips in my hands.

When we came inside the aviary, he would hop all over us, sit on a shoulder, and would cry, "Pete, Pete, Pete, Pete, Pete, Pete, my name is Pete, my name is Pete!" Nick had taught this

Pete the magpie.

phrase to Pete, and Pete loved the sound of his own voice.

By summer's end, Pete was well settled in. Eric was off to graduate classes at the University of Arizona. I was looking for a job, so spent my days on the telephone, as well as puttering around the hills overlooking the small cluster of rental houses where we lived, learning the new set of plants and animals. We had only one car at the beginning of that first autumn in Tucson, so during most weekdays, I was home.

Our small circle of pale-brick rental homes, isolated in a dry gulch, was very quiet, since everyone but me was somewhere else during the day.

One afternoon I was folding laundry in the bedroom when I heard a sound that was unmistakable—the sound of sliding glass doors opening. Was someone else in the house? I felt the hairs rising on the back of my neck as I tiptoed out to the kitchen. No one. No one was in the house. I shoved the sliding door open and looked out into the carport. Jauntily, Pete hopped from branch to branch in the aviary. He looked undisturbed. "Pete," he said in a conversational tone when he saw me. "My name is Pete."

I went back to the bedroom, scratching my head.

I had just finished folding the clothes and was thinking about making myself a sandwich when I heard another sound, a sound that made my stomach clench. Someone had pushed back a chair from the kitchen table!

This time I ran to the kitchen. Someone *was* inside the house!

No one. Through the glass doors I could see Pete taking a drink from his water pan.

I searched the house, even looking in the closets and under the bed. Nothing.

Then I heard the sliding door move again. Again, I ran to the kitchen. Was this tiny house haunted? Did I have a ghost?

Just as I got to the kitchen, I heard the sliding door move again, and again. Then I heard a chair pushing back from the kitchen table. Yes, it was Pete. He was flipping from branch to branch in the aviary, moving chairs and opening and shutting sliding glass doors with that clever magpie voice. I had forgotten that birds can mimic many sounds, not just human voices. What had Pete been hearing? Chairs pushing back from the kitchen table and the sliding glass doors opening and closing, that's what.

Soon I realized that Pete had quite a repertoire. Most of the sounds of barking dogs in our neighborhood were Pete. He perfectly imitated the sound of our small station wagon coming down the dirt road to the house. He imitated the curve-billed thrashers, the flickers, the *burrrr* of the cactus wrens. Perhaps his best feat was mimicking a conversation between two people, sounding just enough distant that the words couldn't quite be distinguished. Countless times I went outside, thinking there must be people standing in the front yard talking, even after I knew that Pete could do this.

I think Pete rather spoiled my enjoyment of other birds for a while, at least of some other birds. For about a month, Pete also hosted a guest in his aviary.

This was a beautiful Mexican jay, about the same size as Pete, but lacking the long tail and clad in a feather coat of tasteful dusty-blue and gray. One of the other graduate students had captured this fellow in a mist net in Madera Canyon after he had noticed that a misaligned beak was keeping the jay from eating properly. We offered the jay a place in Pete's aviary for a while, so that the bird could be hand-fed to gain strength.

The professor who planned to correct the jay's beak wanted to make sure that the fellow was strong enough to withstand the anesthetic of the operation.

Pete tolerated the jay. The only time he showed any aggression was when a dish of raw liver was offered. I would stand inside the aviary and feed Pete liver strips by hand while the jay, behind us, took his portion from a plate.

The jay was beautiful, but compared to Pete, he was boring. I named the jay Week. That's what he said: "Week." All day long, every day, Week said, "Week, week, week, week, week, week, week."

By this time, Pete had learned many more phrases and odd sounds that he would demonstrate often. "What's your name?" he would say, cocking his head to one side and snapping his beady black eyes. "My name is Pete! Hello, Pete! Give me my liver! Got any liver? Is Pete hungry? Pete is hungry! Let's go for a ride! Want to go for a ride? Look, Pete! Stop that! Shut up, Pete." –and so on. Pete imitated silverware clinking in the sink, drawers being opened and shut, the ring of the telephone, doors slamming, and even the hiss of the teakettle on the stove. But Week said only "Week." So, I was pleased when Week, having spent, yes—another week with us after his operation, to make sure he was doing well—was taken back to Madera Canyon and released.

Pete's vocabulary increased greatly by mimicking dog obedience commands, since the patch of lawn where I worked with our Shelties and kept their jumps set up, was next to the carport aviary. Pete watched us intently and listened carefully to the repeated commands. He learned them all.

By the hour, Pete would call out brassily, "Duncan, down! Good boy, good boy, you are a good boy. Over! Over! Fetch. Sit. Duncan, heel. Heel. Down. Stay. Heather, come! Take it. Take it. Good girl, good girl." It often sounded as if Pete were holding an obedience trial all by himself.

Before long, I had a car of my own, a used green VW bug given to me by Gramps and Grandma Hicks, and I could go job hunting in earnest. I soon found a job, a good one, so my long weekdays with Pete came to an end, and every morning when my husband took off for the university, I climbed into my green bug, now named The Desert Frog, and headed down River Road to the U.S. Department of Agriculture Bee Research Laboratory, where I was now a scientific illustrator.

Our tribe of Shelties was growing, and we bought our first show puppy, a blue female named Ballad. Pete was soon rattling off her name as well. "Ballad, come. Look, Ballad, look!" At times he sounded like a first grader reading.

But then came the storm.

One day while I was at work, a late-fall thunderstorm ripped across Tucson and its desert foothills. The storm dropped little rain, but high winds did considerable damage. Trees toppled, fences shredded, screen doors sailed away down driveways, and children's toys skidded along residential streets by the thousands.

That evening when I got home from work, I found my husband standing in a ruined aviary. The dead juniper tree inside had been blown against the door, and the door hung open and twisted. One hinge had pulled free of the frame. And Pete was gone.

With flashlights, we scoured the neighborhood that night until long after dark, calling for Pete. The gulch was home to many coyotes, not to mention owls, hawks, feral cats, and other hazards.

We spent the weekend searching the draws and hills along River Road, calling and calling. Pete would be quite visible, we thought, with his sharp black and white coat of feathers and his long black tail. He would be the only magpie for two hundred miles in any direction. We knocked on doors. We asked people on the street. Had they seen a cocky black and white bird with a long tail? Had they heard a bird talking? No one had.

We searched for Pete on and off for weeks, well into the winter months. Several of the other graduate students, a few of whom were avid birdwatchers, came to help on weekends. We never found Pete.

We repaired the aviary and played host to several more jays for a while, but it wasn't the same.

Our Sheltie family, we felt, was a bit small, but we couldn't have more Shelties in the rental house, so we began to look for property out of town.

At last, we found a good place—ten acres northwest of Tucson near Saguaro National Monument (now National Park). We had a mobile home placed there and began gearing up to move. We wouldn't have electricity there for several months, and wouldn't have running water for a long time after that, but we didn't care. The place would be our own, and we could have Sheltie puppies there. But, what about the aviary? Should we move it to the Tucson Mountain acreage? Would we ever see Pete again?

Sometimes, truth truly is stranger than fiction.

One Saturday in early spring, as we were packing boxes to be moved to the new place, we got a call from one of the other graduate students, one of the fellows who had helped us in the search for Pete.

"I think I know where your magpie is," Linn said over the phone.

We were speechless.

"My wife got her hair done today," Linn went on. "My wife's hairdresser is a very chatty woman. I guess she goes on and on about her clients all the time, and her clients talk to her as well. Today she told Bonnie that one of her clients, a wealthy woman who lives in the foothills above River Road, had found a magpie. She found him after that big storm we had last fall. And get this—she has aviaries. She keeps hundreds of exotic birds there. When she found this magpie, he was trying to get into an aviary."

My husband said, "Do you have this woman's address?"

And Linn said, "I sure do. I want to go with you. I'll come pick you up as soon as I can get to your house."

The woman's home was less than two miles west of our little rental house. Her aviaries were enormous, lush with tropical plants and bright with birds, mostly parrots.

The owner was very skeptical that a university student in worn jeans and faded t-shirt could be the owner of such a splendid creature as her new magpie, but she agreed to let the two graduate students into the aviary where the bird was living.

"My magpie isn't banded," she said as she unlocked the door to the huge, two-story aviary. "I don't know how you are going to prove that he is yours. You are not taking him unless you can prove ownership."

As my husband and Linn entered the aviary, the woman stood to one side frowning, her hands on her hips. Birds screeched and flapped overhead, huge macaws in red, green, and blue, all in splendid condition. No magpie was to be seen.

Then Eric called out, "Pete?"

Suddenly a small black and white whirlwind appeared, cawing wildly. This creature perched on the shoulder of that faded t-shirt, grasping the fabric tightly in its black claws.

"He's very friendly," the woman said coldly. "He likes to sit on everyone's shoulder. That proves nothing."

Then Pete took matters into his own hands. "Pete!" he screamed. "Pete! Pete! My name is Pete! Pete is hungry. Pete wants to go for a ride! Look, Pete, liver! Down. Stay. Duncan, fetch! Good boy! Pete, Pete, Pete!"

Reluctantly, the woman cracked a smile. "I guess he is going home with you," she said.

"Guess so," Linn told her.

And a few minutes later, Pete was home in his aviary, happily jumping from branch to branch in his juniper tree, opening and closing those doors and drawers, hissing like a teakettle, and shrieking "Pete! Pete! My name is Pete!" until I thought his throat would burst.

We moved the aviary to the new place and set it just outside the front door of the mobile home, next to the yard where we did our obedience training. Pete learned new commands along with the Shelties. Life was good.

One morning I came out into the early sunshine and sat on the front steps to put on my shoes. The sky was bright and cloudless, and from somewhere nearby I could hear a cactus wren calling. But something was odd. Something about the day struck me as strange. I couldn't quite put my finger on it.

Then I realized. It was Pete. Where was Pete? Where was our guy, chattering and making those door-sliding noises, hopping madly from branch to branch as soon as he saw me coming out the door, hoping I had some liver?

I found him dead on the sand. Lying next to Pete was the body of a huge toad. I knew the species immediately, *Bufo alvarius*, the Colorado River toad, dangerous and toxic, poison seeping from its glands and skin. The toad, perhaps attracted to Pete's water dish, had dug its way into the aviary from the bottom, and Pete had killed it, but not before getting a lethal dose of the toad's poison himself—a sad and sudden end to the bright and raucous spirit that had been part of our lives for the past four years.

Back in Idaho after many years, I see magpies every day, scratching under the bird feeders and hopping with alacrity from branch to branch in the cottonwood trees in the bottom of our gulch. I see pairs of magpies, claws holding tightly to the top strands of barbed-wire fences as they face into the brisk wind of another bright morning, and I think of Pete. "Hello, Pete," the wind whispers. "My name is Pete! Duncan, come. Heather, down. Look, Ballad, look. Pete is hungry. Got any liver?"

The Halloween Costume

I was throwing away snakes and other creatures from the small teaching collection of the Biology Department at Millsaps College in Jackson, Mississippi, and I was glum.

Eric, now a professor there in his first postgraduate position, needed more room for other things. In this collection were a number of specimens decades old, in many cases poorly preserved and without documentation—that is, unlabeled. Most biological specimens are valueless if there is no information with them that notes where and when they were collected, and who did the collecting.

The contents of these dusty old jars hadn't been used for research or for classes in decades—probably had never been used at all. I found countless jars of crayfish (some unnecessarily and boldly labeled "CRAWDADS") in which the preserving fluid had long dried up; several jars of earthworms so rotten that even a catfish would turn up his nose at them; jars of snails that stank disgustingly; a jar of salamanders with no tails (why?), and countless jars containing various other unlabeled creatures. The jars themselves were non-standard, ranging from Mason jars to ketchup bottles, and even a squat blue-glass Vicks jar that held a tiny rabbit foot. Most of these specimens were destined for the fifty-gallon garbage can I had dragged up the stairs from the greenhouse that morning.

Then there were the snakes. Someone, at some distant time (probably in the 1930s), had collected hundreds upon hundreds of tiny, newborn-sized common garter snakes. The garter snakes were coiled into mayo jars, Mason jars, pickle jars, and olive jars. Their slim little tails at the bottoms and their tiny bead-eyed heads looked blankly at me from near the tops of the glass containers. Who would collect over a thousand baby garter snakes? They were too small to use for teaching the reptile organ systems to a zoology class. They looked as uniform as if they had been stamped from plastic in a toy factory. *Poor little things*, I thought, dumping a jarful into the garbage can.

I was also feeling sorry for myself that day.

I had been invited to a Halloween costume party. I love Halloween, and no one had invited me to a costume party ever before. The hosts were a sweet couple, students at the college, Tim and Mary, engaged to be married in a few months. My husband was going as Dr. Jekyll, which name I had embroidered on one of his white lab coats. But what would I wear? The party was to begin in little more than 24 hours, and I had no costume. I dumped more snakes.

Near the end of the month in those tight times, we were out of funds for anything but the bare necessities. I was never out of ideas, but kept thinking to myself, *the only time* ever *that I've been invited to a costume party—a* Halloween *costume party— and I can't buy anything to make myself a great costume.* There would be food, there would be bobbing for apples, there would be ghost stories, there would be good friends, and there would be music and dancing. I sighed and reached for another jar of snakes.

I can't dance worth beans, have never been able to, but for some strange reason, I wanted to go to this party and dance. Sort of. I stared into the nearby lab sink with one hand holding a jar of sunfish. These were labeled properly, so I could wash the dust off the jar and put the sunfish on the keeper shelf.

Cinderella-like, I said to myself, "What would your ideal costume be, if you could have anything your heart desired?" Surrounded not by helpful little mice and birds, but by jars of dead snakes, salamanders, and fish, I thought, *Well, I'd want to be classic and regal. I'm 36—I'm too old for something cute. I'd want a costume that would make people want to ask me to dance, but, but—something that would make them leave before they found out that I can't dance.*

I don't do closeness very well, either. I reached for another one of the olive jars full of little snakes.

As I was turning the jar to look for a label, I noticed that the glass was cracked and leaking. I poured out the fluid. Two of the snakes came out as well, frozen in ringlet loops after their long stay in the tall, narrow olive jar.

A light bulb came on. I gathered an armload of olive jars filled with baby garter snakes, wrenched open the lids, and shook all the snakes out into the enormous laboratory sink. This was the beginning of something interesting!

Each of the little snakes was frozen into a long, loopy curl. But the snakes stank of formaldehyde, enough to make my eyes water. That wouldn't do. Then I noticed a bottle of dishwashing liquid on the counter. Yep. I'd do snake washing.

After rinsing the curly snakes under the tap, I filled the lab sink with hot, soapy water, enough to cover the snakes, and left them overnight.

First thing the next morning, I rinsed off my snakes and took a long sniff. No odor! Bundling them into a box, I took them home.

Out from the linen closet came a white sheet and from the desk drawer, a box of big safety pins. The sheet, strategically pinned, would be my gown.

From a strip torn from fabric in my remnant-basket I made a white headband and centered it with an old white-rhinestone starburst pin that was once my mom's. Then I took some clear fishing line, a big needle, and started in on the snakes.

Oh, yes. You have figured it out by now. I would go to the Halloween party as Medusa.

Using clear fishing line of varying lengths, I hung about two dozen snakes from the back half of the headband, some snakes hanging by the head, some hanging by the tail, all looking like little sausage curls from their long stint coiled in the olive jars. Their intricate scale patterns gleamed in the light over my worktable, and their polished black eyes shone.

Then with sheet and safety pins, I made a draped one-shoulder gown in the Greek-statue fashion and finished off the costume with a pair of leather sandals. Half an hour before we were to leave for Tim and Mary's party, I got out the curling iron and turned my waist-length dark-blonde hair into long sausage curls.

Hubby donned his Dr. Jekyll lab coat, I bound on the dangly-snake headband and distributed the garter snakes among my curls, and in the dark of the moon, we drove to town for the Halloween party.

The dancing had already begun. Right away, one of the students, Joel, asked me to dance. We began. I tried to dance. Joel said, "Wow, that is some costume, Dana. Those snakes look really real." I saw his hand reach out toward one. "Aaaaah," he screamed. And Joel was gone.

Next to ask me to dance was Jeff, one of the lab assistants. "Wow," he said, swaying with me to the "Monster Mash." "Where did you get those rubber snakes? They must have been really expensive, because they look—" here Jeff reached for one. "Augggh! They *are* real." And Jeff was gone.

Then, Professor Bob came and departed abruptly ("Holy crap!" was his reaction.) and then Kelvin ("Noooo!"), and—ah, I had a fine time almost-dancing that night.

By midnight, when the child-sized coffin was carried into the living room, the book of ghost stories taken from it, and the tales begun, no one would sit near me.

This was the only costume party I have ever attended. Though it happened so many years ago, I have never gone to another.

It's not easy being a monster.

The Creature from the Small Lagoon

One detached side of a small slat-wood crate hangs on the kitchen wall near our back door, where it has rested for many years. The wood is old and sports salt stains and the circular footprints of tiny barnacles. It has been a fixture in many kitchens where I have lived, and to me this square of slat-wood is a painting of a sort, a strange rectangle of parallel laths, splintery wood, rusted nails, and a few empty nail holes. The wood no longer smells of the sea, but still the sea calls to me from its space on the wall as I go in and out the kitchen door to water my pots of marigolds or to play with the dogs.

In the mid-1970s, we led a group of students from Millsaps College on a winter for-credit field trip to southern Florida. We hit the road in late December, just after Christmas.

We had been on such a winter Florida expedition once before and had visited the Keys, staying in my uncle's cabin on Big Pine Key. By this time, however, the cabin had been sold. On this trip, we planned to find a campground of some sort on or near Big Pine Key, if we could.

As luck would have it, the day we arrived in the Keys, we spent the afternoon near the visitor center for Key Deer National Wildlife Refuge, exploring a small area of native vegetation on Big Pine Key. One of the students brushed against a poisonwood tree and broke into a painful rash on both her forearms. We sought help from the station director, a famous conservationist (let's call him John), who broke a fat leaf from one of his huge outdoor plants of aloe vera and rubbed the cool sap onto Gail's angry welts to soothe her pain.

We hit it off with this fellow and his wife and stayed to talk with them for some time. Suddenly, John made us an offer we could not refuse. "Why don't you and your students camp on the refuge tonight?" he asked. We knew that camping on the refuge was prohibited. "I am convinced that you won't harm anything," he continued. "I'll direct you to a place that is very private, a nearly enclosed cove where no one can see you and no one will bother you. There's some driftwood on the beach there. You can even have a fire if you keep it small."

We were sold. Within a few minutes, we had reached John's private cove by bumping along a tiny, almost invisible road made of coral and shell rubble.

We had learned from the director that this area had been slated for development several years before. Plans for building had been drawn up and approved. The area surrounding the small cove was to be turned into a place of exclusive waterfront homes, each of which was to have a deep slot or slip where a large boat could be moored. A channel was dredged and blasted through the cove all the way in from the sea, with branches leading to each of the proposed homesites. Each homeowner would be able to pilot a deep-draft yacht from the open sea into the cove and moor it to a dock attached to his own home.

Then, Key Deer National Wildlife Refuge was created, stopping the planned development in its tracks. But by then, the channels in the shallow cove had been blasted and dredged deep below the surface, as had the individual mooring slots all around the edges of the cove, one for each platted homesite. But that was as far as the development progressed. No homes were built there. The place was still wild, a tangle of low scrub, tough grasses, red mangroves, and woody vines, home to the small deer of the Keys. The placid surface of the water gave no hint of the dark channels that had been created below the surface.

The cove, with a very narrow opening to the sea, was its own tiny world. Few people had come to the cove since the refuge had been established. No one had spent a night at the cove for years.

We pulled our van up to the beach of coral sand and made camp, gathering driftwood as the early darkness of midwinter fell upon us.

The night was brilliant and cloudless, with millions of unblinking stars. A slight chill shivered across the black water, and we were glad to light the green enamel lantern and matching Coleman stove, preparing to cook ourselves a hot meal. The coffee pot was filled and placed carefully on the fire.

Before long we were talking merrily around our campfire as we sat on the clean sand of the cove eating our meal. The flames leaped and reflected in the dark waters, and an early moon rose nearly full over the palmetto and satinwood scrub.

At a great distance through low trees, we saw a wink or two of lights from houses elsewhere on the Key, but thanks to John, we had achieved the impossible for such poor interlopers—a private beach in the Florida Keys during the holiday season.

A light wind sprang up and we drew closer to the fire, hearing the *slap, slap, slap* of wavelets in the mangrove roots as the tide began to come in. The moon shone clearly from halfway across the sky, paling the stars and painting a trail of rippled silver far across the cove.

I studied the firelit faces of the students. Some heads were nodding already. This had been a long day for them, and they were very tired.

I, however, did not feel sleepy. Tonight, I would do the dinner dishes as the students dozed over a final cup of coffee. I would not use soap in the cove but would scour the plates and silverware with coral sand and rinse them clean. If necessary, I could wash them with soap the following day in a better place to use detergent.

I picked up the Coleman lantern and pumped it until I could hear it hiss. I carried the lantern by slipping its bail over one arm, so I could also grab double handfuls of dirty silverware and take it all down to the water's edge at a little distance from our fire.

I set the lantern on the damp sand within inches of the water. The place where I knelt to the tiny inland sea was strange, a very narrow finger of land jutting out into the cove, a tiny peninsula shaped like greater Florida itself, but not as wide as the length of my own arm.

Laying the dirty silver on the sand, I began to scrub the forks, knives, and spoons one by one, polishing them with care. I was in no hurry tonight.

To my left, the lantern light shone through ten inches of sea water into the shallow cove, a huge private aquarium where countless thousands of upside-down six-inch jellyfish, looking like miniature cancan petticoats, lay pulsing on the bottom, a living carpet extending as far as the light would reach, and even farther. On my right and nearly at my toes, a void opened under the water, fathomless to the eye, deeper than deep, a great and sudden blackness: the dredged slip for a boat that would never come. In that blackness, something moved.

I stiffened and dropped a knife onto the sand. Rising from the deep came two dinner plates. Startled, I stood and watched them surface, then laughed aloud.

The dinner plates were horseshoe crabs, those pale, ancient sea creatures older than the dinosaurs and not like true crabs at all, with smooth carapaces the color of creamy jade and the world's strangest eyes. A male and female, they clung together in the age-old dance. The moon had called them. By morning there would be egg cases.

I knelt again, cleaning the silverware fork by fork, knife by knife, spoon by spoon, moving each piece from the soiled pile to my cleaned pile, singing softly to myself. The only other sound was the quiet lapping of the waves. The black sky formed the top of the world, with the silver moon melting into the cove below, looking down on all us creatures as it always has, creatures ancient and modern, time-worn and crisply new.

Snatches of sleepy conversation came to me over my shoulder. The students were drifting away from the fire, bound for their sleeping bags. Once in a while, I could hear the pop and crackle of the fire itself; fires made from driftwood always burn with little surprises.

No cars roared, no radios blared, no televisions flickered. The cove smelled of the sea. The moon trail tangled in black mangrove roots. Not a soul was to be seen abroad in the night, loud or stealthy. It was as peaceful a camp as I had known, and I had known a great many camps.

Finally finished, I gathered up all the utensils. Kneeling, I sluiced the two-handed bundle of cleaned silverware back and forth in the shallow water above the resting jellyfish to remove the last of the sand.

Something rocketed from the black depths of the boat slip and smashed into my hands, scattering silverware everywhere with a ringing crash.

Stunned, I watched an arrowed wake departing at speed, a pointed fin and tail cutting the surface of the shallows. Flashes of mirrored scales glinted in the wake. Another heartbeat and it was gone.

Silverware lay gleaming like strange jackstraws across curled jellyfish. *What?* I thought belatedly. *What?* The sea rocked, smoothing.

Whatever had come, was now hidden in the central channel. Far out in the cove, a tiny star blinked on the edge of a wave, then vanished.

What? I thought again. But by then I knew: the fin, the tail, the bright scales, the speed. And I thought of the flashing and chiming of the silver.

Into my gentle idyll had surfaced a barracuda, nicely at home in the depths of the blasted boat slip. The cove was no longer simply a shallow inlet fit only for harmless jellyfish, crabs, and snails. The developer had invited to this once-calm place the depths of the open sea. I had not considered that. Indeed, heedless, I had done precisely what I had told the students never to do in Florida's seas: flash metallic objects at the surface, things that might look to a barracuda like a school of silvery minnows.

The cove barracuda had slashed through the "minnows" as his kind often do. I was lucky to have all my fingers. Sobered, and with a wary eye to the black depths of the boat slip, I lifted each piece of silver from the pillowy and perhaps astonished jellyfish, took up the lantern, and went back to camp.

On the way, I stumbled over the slat-wood crate.

Today, one side of the little crate hangs on the kitchen wall, a found object that looks strange in my Idaho kitchen. Each time I notice it, I hear the tinkle of falling silver, see the flash of mirrored scales in the dark, and remember to be wary in all the quiet places of the world.

Richard Nixon

He had a distinctive profile, especially the portion that included his nose. He was short-sighted. He spent a lot of time covering up. Richard Nixon was a mole.

When I lived in Jackson, Mississippi, I was often out and about turning over logs, looking for salamanders. I have always loved them—the puppy-dogs of the amphibian world—with their bright dark eyes and long waggy tails. My then-husband and I would go hunting for small creatures to showcase in a terrarium in one of the labs he taught at Millsaps College, leaving them in a lab for a few weeks before we released them.

One day in early spring, we were turning over logs and searching in heaps of damp leaves at the upper end of Ross Barnett Reservoir, hoping to find some spotted salamanders. It was a good time to hunt for them; rain had soaked the area for the past several days, and everything was damp.

It soon became apparent that the reservoir water was rising. We were some distance from our car when we realized that if we didn't hurry, we'd be wading in ankle-deep water before we got back to where we had parked our car along a small dirt road. During our hunt, we found two spotted salamanders, dropped them into one of the cloth specimen bags we carried, and hurried on.

Pretty soon we ended up wading anyway, and it was a relief when we came to a small island in the rising water. It was about fifty feet in diameter, topped with a short length of rotten log. We could tell that the little island would be under water within the hour, but I couldn't resist rolling that rotten log.

Under it, in a topless tunnel, huddled a small brown ball of fur. I scooped the small one into another collecting bag, figuring that if I didn't take the creature, it would likely drown.

The creature was a mole.

Moles, like shrews, are insectivores, and they have an incredibly high metabolism, requiring food every few hours or they will starve. They eat all kinds of larvae and worms, creating shallow tunnels in the soil with their paddlelike feet and tough claws. Their eyes are tiny and barely functional, but their noses are fabulous. Many people think that moles are the same as gophers, but no! Gophers are rodents, with MUCH more functional eyes and a rodent's rather ordinary metabolism.

Once home, I dove into our laundry room to get my old aquarium. I'd had it since I was a little girl in Idaho. The tank, made by a cook at Sun Valley Resort, had been fashioned from automobile window glass gleaned from a wrecking yard in Twin Falls, and when it was new, held nearly thirty gallons of water. I had saved up and bought it from the cook when I was twelve, and though the tank had sprung a leak after one of our many moves, I hadn't been able to make myself throw it away. The tank would no longer hold water, but it would hold dirt.

I half-filled the tank with dirt from the backyard, installed a water dish and a couple of handfuls of grass, fashioned a cover from a piece of hardware cloth, and dropped in Richard Nixon. True to his Watergate nature, he immediately dug himself a hole and disappeared. Shovel in hand, I headed again into the backyard, this time in search of worms.

Richard spent most of his time underground, so you might wonder why I found him such an interesting and charming pet wasn't only for his delightful personality.

The first thing he did was to dig himself a den. Lucky for me, one side of his den was glass, so I could look in and watch him sleeping, his little cheeks huffing in and out, his legs twitching. Occasionally, I would hear tiny creaks and squeaks as he snored. I soon exhausted our backyard of worms, but I found that I could buy mealworms for Richard at the pet store and nightcrawlers for him at bait shops. Our fridge was always stocked with live worms.

I loved to watch Richard feeding. I'd drop a long worm onto

the dirt in his tank. Soon, perhaps because of the movements of the worm carrying through the soil, Richard would wake. Then, watching from the top, I'd see a flat pink nose poke up from the dirt. This nose would appear near the worm and go patting around until the scent was well and truly caught. Then Richard's head would erupt from the dirt and smack down on the worm. He'd sit up like a short squirrel and, bit by bit, feed the worm down his throat. Then, he would groom himself with those big-paddle paws and go back to his den. I fed Richard six to eight times a day. He was always hungry.

Richard was at his most entertaining when he was taking a bath.

One day I thought the soil in his tank was getting a bit too dry. I had a small watering can I used for house plants, and from its narrow spout I dripped water onto the dry dirt of his tank. Richard charged from the den and stood under the dribbling water. Squeaking, he took a shower.

He rubbed himself all over with those huge paws until his fur was thoroughly soaked, turning around and around, chirping and creaking. I could swear he was singing. When I stopped pouring water, he would spend some time grooming his fur, and then he would go below for another nap. After that first time, I gave Richard a bath every day.

But then the calendar rolled around to autumn and, one by one, the bait shops closed. One day the pet store had no mealworms. I had dug our yard pretty much free of worms. Reluctantly, I took Richard out into the forest, to a place much like the area where I had found him and let him go.

Did Richard give a backward glance? No. In the space of ten seconds, he had dug himself out of sight.

Too Close

F *lashback:* It was the second summer after I graduated from college, 1967, and I was teaching high school in Oregon, while my new husband was in the process of getting his master's degree in entomology from Oregon State University. To complete his thesis, he had to travel to several entomology museums and university collections, to compare his beetles with the specimens collected by others, the "type" specimens that defined each species when they were originally described. The type specimens Eric needed to see were located at places scattered across the continent, including Purdue University in Indiana, LaValle University in Quebec, Philadelphia Academy of Sciences, the American Museum of Natural History in New York, the Smithsonian in Washington, D.C., and Duke University in North Carolina. Armed with a good camera, reference specimens, and drawing pads, we set out.

My younger sister Vicki, and Eric's youngest brother Bob had just graduated from high school, and I thought it would be fun to take them on our cross-country trip. We traveled with a station wagon, a little teardrop trailer, and a tent. In a couple of locations, we had friends who invited us to stay with them, but mostly we camped—often fireless, waterless camps where we would pull over on deserted roads in the middle of the night after having spent a day poring over specimens in some museum. We were careful and lucky, not often having to pay to spend the night in actual campgrounds—and we were usually not noticed. Often, we would be back on the road at dawn before anyone in the immediate area was stirring.

Things were going smoothly until we got to Boston.

We spent a very long day, and most of the twilight, working in the entomological collection at Harvard University.

That evening all four of us were very tired, Eric from staring through a microscope for many hours, and the rest of us from spending those same hours looking at everything we could see in the Museum of Comparative Zoology, marveling at the progression of Simpson's original fossil horses—from little fellows about the size of pygmy goats to close ancestors of the horses we know today—plus the many glass models of certain plants, and the most gorgeous beetles we had ever seen. We peered into cases of butterflies and trilobites and countless other treasures, and almost on tiptoes, walked past the offices of great modern scientists like Alfred Romer and P. J. Darlington.

And then, in the deep dusk, we got into our station wagon and took off. The next day's destination was the American Museum of Natural History in New York City. We needed to get out of Boston and its suburbs and find a place to camp so we could get some sleep so we could be on the road first thing in the morning.

Darkness found us in Connecticut, on a tree-lined highway in some little hills along the Housatonic River. This seemed to be a sleepy, pleasant area, so we looked for a gravel road with a pull-out, a wide spot—and near the top of a small hill above the river, we found one.

It was perfect. No one was around, and we could drive completely off the road onto a grassy place under a big tree. We were so tired that cooking dinner was out of the question. The night was clear and there didn't seem to be any mosquitoes, so three of us rolled out the tarp and flopped sleeping bags down. The night was warm; Vicki, Bob, and Eric lay down on their bags.

I was the holdout. Irrationally, I felt that something was wrong. The others begged me to get out of the car and get some sleep. "This isn't a safe place," I insisted, not knowing why I was insisting. "We have to leave." I stayed in the car, sitting bolt upright.

The others gave me a "Fine, do what you want, but you are crazy, and we are going to sleep" speech, and lay out on their bags. But I couldn't sleep. Something was wrong.

Before long, very faintly, I heard screaming from below, down by the river. Over the course of a few minutes, the cries grew louder. Suddenly I could see flashlight beams dancing through the trees. I saw the forms of several people, moving shadows. After a while, I realized that they were coming closer, coming up the hill toward our camp.

Who were these people?

Some were screaming, some were laughing like maniacs, some were shouting. I remembered the binoculars stashed in the center console and grabbed them. The glimpses of faces I caught as the group moved uphill through the trees showed me young people, college-age, perhaps 20 to 30 of them. In the erratic flashlight beams, I saw that some of the people had syringes and were stopping briefly to use them.

Drugs, I thought. I had heard about such things. I watched in fascination as people bashed into trees, fell into bushes, picked themselves up, and kept coming. It was like watching an invasion of zombies.

I will never forget the glancing flashlight beams, the staggering steps, the slack arms hanging down as if useless, and the insane laughter. The back of my neck prickled, and my hand reached for the door handle. We had to get out of here, now. I had to wake the others.

But as I began to open the door, there was my sister beside the car, sleeping bag bundled in her arms. By this time Bob, Vicki, and Eric had also heard the screams and had seen what was coming up the hill.

Shrill laugher rang out. The mob was closing in.

Sleeping bags and tarp were shoved into the back of the station wagon, and we took off just as the first of the strange crowd reached our grassy flat.

We drove for another hour and finally spent the small remainder of the night sleeping in the car at a rest stop.

After that, through the years, Eric and I made many camps in many places, hundreds upon hundreds of camps, both on expeditions and on our own. And every time we stopped, Eric would ask me, "Is it safe to stay here?" And I would try to guess, to feel, and not really understanding, but not feeling anything dangerous—I would say, "Yes, this place is safe." Whether it was a national park campground in the swamps of Florida, an unnamed gulch in British Columbia, or a canyon in our own familiar Idaho, I would take a few moments to breathe, and then I would tell him if it was safe.

Then.

Fast-forward to July 31, 1976, after Eric had been a biology professor at Millsaps College for several years. Having "grown up" on college biology field expeditions, one reason we had been attracted to Millsaps was its field biology program, which Eric was happy to continue.

That summer, we traveled in an eleven-passenger Dodge van with six students and Duncan the Sheltie. Pulling a small trailer crammed with a camp stove, pots and pans, tents, sleeping bags, personal duffels, a library of books, several plant presses, and other implements of field biology, we took off for Rocky Mountain National Park in Colorado. We spent the summer lecturing, keying out plants and animals, live trapping and releasing small mammals (with a permit, outside the park), cooking over a fire, doing ecological transects above timberline, flyfishing, getting rained on, going to town, putting our sleeping bags in the driers of a laundromat, coming back to camp, getting rained on, going to the laundromat, coming back to camp. And getting rained on.

We had a good group that summer, a handful of intelligent, witty, and interested students (one of whom is still a dear friend). We survived having a baby porcupine chewing on the underside of the van, Duncan getting his paws blistered while walking on hot sand at Great Sand Dunes National Monument, a lost contact lens, and other small mishaps. But all in all, we had a good trip.

Then, it was time to break up camp and head from the Rockies back to Mississippi. Of course, it was raining. We left our primitive camp in the Mummy Mountains, swung up over Trail Ridge, swooped down into the deep valley to the east, and said goodbye to the Rockies.

By the time we drove through the little town of Estes Park at the end of that day, we were beyond tired, and we were hungry. The sky was overcast, but at least it wasn't raining. We drove along the Big Thompson River looking for a place to camp, a place with real bathrooms and picnic tables, for our last night in Colorado.

Eric slowed the van. "Look," he said, "that's a campground, and there are some spaces free, too." He turned to look at me. "There's the access road. What do you think?"

And without thinking, without considering at all, I said, "I don't like it." The students were not happy.

We continued down the narrow, winding canyon and came to another campground, and another. By this time, we were overtired and getting a little whiny, all of us. "I don't like it," I said again. The students watched as the clean, lovely campground— with its restrooms, tables, tall pines, joyful little river, and level places for pitching tents—disappeared behind us. We moved on.

We came to the entrance of the last campground before the land would open out into flat, featureless farmland and scattered-ranchette suburbia and I said, "I don't like it," once more. Eric glanced at me, irritated. Then his face changed, and he said, "Connecticut?"

I nodded, feeling goosebumps breaking out on my arms. "It's not safe," I whispered. Subliminal cues were speaking to me all at once, in silent voices.

It was nearly dark by the time we found a place to camp in the flat farmland somewhere east of Loveland: a little-used dirt road down a slight slope and at the end, a fairly level, weedy area at the bottom of a sugar-beet field. It was ugly and could not compare to the lovely pine-sheltered campgrounds we had passed by, but as I got out of the van and took a deep breath, I relaxed. Ugly it might be, but this place felt *safe*.

Then I took a step and sank into three-inch deep mud. At least it wasn't raining, yet.

We cooked dinner and left our shoes, caked with mud, outside the tents, hoping it wouldn't rain. It rained during the night.

The next morning, we pulled our feet from the sucking mud with each step as we packed up the tents for travel. I looked at the brown-splattered and bedraggled crew and couldn't face trying to cook breakfast in the muck. I did some quick calculations. We had perhaps two nights left on the road. Was there enough money? There was.

"We are going out for breakfast," I announced. The students finished packing with alacrity. I was astonished that the van didn't get stuck in the mud on the way back to the pavement.

We found a diner not far along the highway in Denver, and trooped in. Quickly, we found a table long enough to seat us all and were served steaming coffee.

Our waitress seemed distracted.

Oddly, everyone in the diner seemed distracted. Their attention was riveted to a small TV mounted above the counter. Our waitress returned to take our orders, and as she strode off to the kitchen, I saw her turn up the volume on the TV.

"Dozens are missing," a deep voice blared from the TV. "Highway 34 is closed from Estes Park to Loveland. Phone lines are down. The state patrol has sent troopers into the canyon."

Suddenly our attention, too, was riveted to the TV. I cradled my warm coffee mug in both hands and heard the voice continuing. "If you know of a friend or relative who might have been in Big Thompson Canyon last night, call this number immediately. Helicopter rescue is in progress. Some people have been rescued. We will be posting lists of the rescued and of the dead as information reaches us, so stay tuned."

The waitress brought our food and found us silent, looking from face to face. I don't think I have ever been so grateful to be eating breakfast with a group of students around me, before or since.

Rain in the mountains had dumped more than 12 inches of water into that canyon in four hours, and the river had turned from a rocky stream two to five feet deep, into a raging 20-foot wall of water that had filled the entire bottom of the canyon. The lovely campgrounds were gone. The highway we had traveled was gone.

The toll from the Big Thompson flood of 1976 was 144 dead, with five people missing, their bodies never found. Many of the dead had camped in the very campgrounds we had passed by on our way to the muddy weed patch at the bottom of the beet field east of Loveland.

Ants at *Estero Morua*

Yesterday I came across a photo of Estero Morua, Morua Estuary, a small bay in Sonora, Mexico, tiny window on the Pacific Ocean.

In the mid-70s, my brother-in-law Nick was a graduate student at the University of Arizona in Tucson. While visiting us in Mississippi, Nick mentioned that while he had worked on another project, he had seen ants in this estuary, far down in the intertidal zone. He thought it possible that colonies of ants, not just stray individuals, might be living where they would be covered with two or more feet of seawater for several hours twice a day, when the tide was in. Intertidal ants?

Spring break was already underway (hence Nick's visit), and there was no time to lose. Three people hastily crammed food, water, sleeping bags, buckets, jars, forceps, trowels, cameras, notebooks, Milk-Bones, a Sheltie (Duncan), and a tidal table for the Sea of Cortez into our tiny Mazda GLC hatchback, and took off.

We arrived at the estuary in mid-afternoon of the following day and set up camp in the secluded beach dunes. No one else could be seen. The tide was out, and except for a narrow channel down the center, the estuary was dry as a bone.

We found ant colonies. Two species of ants inhabited the part of the estuary that would be under water at high tide.

We marked hundreds of colonies with flagging.

We ate peanut butter sandwiches as the sun blistered us, and waited for the next tide to come in.

It was fully dark when the tide turned and began to flood the estuary.

Picture a blue merle Sheltie and three people with lanterns, all walking in the water, each person surrounded by a halo of light, our movement causing ripples of gold as we watched the ants retreat into their tunnels.

Banchory Duncan Macduff CDX, HIT, the Sheltie

The tide came in, and the sea covered the colonies, deeper and deeper. Above, the sky was coal black, and the stars were shining brilliantly. There were no lights but our lanterns, no noises but the lapping of water, the hiss of the lanterns, and the howls of distant coyotes.

We measured the water depth at high tide and marked its reach. And the next morning, we sprang up at first light.

The tide had drawn back, and the estuary was dry again. We raced to our marked ant colonies to see if the ants were still alive—and they were. At every colony, the soldiers were already out foraging.

The pattern continued with the next tide, and we collected more data. But spring break was nearly over, and we needed to learn how the ants were surviving underwater for so many hours at a time. This would unfortunately be our last night at Estero Morua.

By that afternoon, we had run out of food. We knew we'd have to drive all the way back to Mississippi in one marathon stretch to make it home before Monday morning classes began, and we had to save our money for gas.

We explored the dark and flooding estuary by flashlight, and discovered that the ants had plugged their entrance tunnels with little bits of clay.

With a trowel, we carefully cut into a number of the ant colonies after the water had covered them. Our lanterns shone down through the clear water and found air bubbles in the walls of the tiny tunnels. The ants clung to the bubbles, the legs of each ant holding a bubble to its underside so she could breathe!

Duncan didn't know why we were hopping with excitement, but he splashed with us in the water and barked as we laughed like maniacs in the dark.

Then barefoot Nick, with a little yelp, discovered that the central channel of the estuary was full of large crabs. He began grabbing them and stuffing them into an old burlap bag he found caught in the weeds at the high tide mark. He harvested several dozen crabs.

By midnight, we had collected all the data we needed for a scientific paper.

We built a driftwood fire to cook the crabs, and exhausted, plopped down on the sand beside the fire as the moon continued to rise above the mountains to the east. We had food!

The only other food we had left was a small bag of limes, so we thought we'd squirt lime juice onto the crab meat for added flavor. But suddenly Duncan stood up and growled.

A dark shape appeared at the top of a dune, coming toward us. Worried, the two men went out to meet whatever it was while I held Duncan back.

The dark shape turned out to be a fellow University of Arizona graduate student, a student that Nick knew. This man had seen our fire and wondered if we might have any food to share. He, too, had run out of food.

All he had left of his own spring break food stash was a bottle of tequila and a pound of butter. Guy with tequila and butter, meet trio with crabs and limes!

After stuffing ourselves, we all sat by the coals of the fire and watched the moon trail through the stars and drown itself in the

Sea of Cortez.

We later found out that one of our species of intertidal ants was a new species, and that we were the very first to document the phenomenon of intertidal ants. Those few days at *Estero Morua* resulted in the first scientific paper of my career (co-authored), and this memory.

The Key

Have you ever used a biological key? No, I don't mean a small metal object used to lock and unlock. For a biologist, a key is the primary way to identify plants and animals to their species. Sometimes, keying out requires a microscope, so you can see structures like tiny hairs. With plants, sometimes you need the leaves, flowers, and seeds to make sure you have made the correct ID.

Biological keys are written identification guides made up of choices, and each choice points to a certain path. Here's an example: 1a: leaves on the stem are opposite one another, or 1b: leaves on the stem are alternate. Let's say the plant has opposite leaves, so 1a is the correct choice. That would lead to 2a: leaves entire (means that they are simple, all in one piece) or 2b: leaves divided (like a fern or an acacia leaf). After many such steps, aided by range maps (no chance that this high-altitude forget-me-not would be found on Key West), with a bit of skill and perhaps a bit of luck, eventually the key-user will have identified the species he or she has just encountered on a trail or a beach, on a mountainside or on a reef. A great part of understanding the real world depends on knowing just exactly what is there!

In the summer of 1977, we decided to do an all-summer, for-credit college field expedition from Millsaps College to arctic Alaska. I bought a hefty plant key in the form of a thick book by the arctic botanist Eric Hulten, titled *Flora of Alaska and Neighboring Territories*. Before we piled into the college van and took off, I spent weeks hauling the Hulten everywhere I went, studying the key and its very detailed drawings of plants that grew in ecosystems I had never seen. I was going to be teaching a field course titled Vegetation of Alaska. I had a decent working knowledge of plant families, so that gave me a head start, but there were a LOT of unique plants in Alaska!

We drove. We drove through Arkansas and Oklahoma, Colorado and Idaho, Montana and Alberta, Yukon Territory, and finally, we entered Alaska. We depended upon the travelers' bible for Alaska, a magazine called *The Milepost*.

We wanted to experience plants and animals above the Arctic Circle—in Arctic National Wildlife Refuge in particular, and another biologist recommended a bush pilot named Walt Audi, who lived on Barter Island (just off the arctic coast) in an Innuit village, Kaktovik. We wrote to Walt and then called him. He agreed to our proposal. He told us which flights to take from Fairbanks to Prudhoe Bay and from Prudhoe to Kaktovik, where we would meet him. In two groups, Walt would ferry us and our stuff about 40 miles into Arctic National Wildlife Refuge not far from the Alaska-Canada border, where stood an abandoned Distant Early Warning military station where we could shelter.

Beaufort Lagoon in Arctic National Wildlife Refuge in July 1977

We couldn't take our Sheltie, Duncan, with us to the high arctic, so we left him at a boarding kennel owned by a Sheltie breeder I knew. Duncan had never been left behind like that before and was highly indignant. And he ignored me for an entire day when we returned.

We drove to the Fairbanks airport and were off!

The sea ice was still crowding the shore on the Fourth of July, a wonder I had never seen before. Far out to sea, we flew over a swimming polar bear; the pilot circled him as the bear hauled itself out onto an iceberg. We got off the plane in Prudhoe, and then boarded a twin-engine Otter for the flight to Kaktovik. When at last we landed on Barter Island, fog had begun to roll in from the sea. The Otter pilot pointed out Walt Audi's house—which, like all the other homes there, looked as if it had been constructed from huge packing crates.

Laden with all our stuff—food, Coleman stove, sleeping bags—we trudged 200 yards and knocked on Walt's door.

"Come on in!" he said with a smile. "You can set up your tents in the driveway and cook in the library."

"What?" I exclaimed. "Aren't we going to fly out?"

"Not today," he said. "Fog."

The library was a small room lined wall-to-wall with shelves filled with paperback books, adjoining a small bathroom. A "small bathroom" was a curtained alcove containing a ten-gallon bucket with a toilet seat on top. When you sat on the bucket, your legs from the knees down stuck out from under the curtain, to be seen by anyone in the library.

Fogbound, we stayed in Kaktovik for three days, and got to experience an Innuit whale festival along with two *National Geographic* photographers who were also marooned there. But that is another story.

Finally, a morning dawned cold and clear, and we were off. I was in the second group that Walt ferried out to the DEW station. The ice had already begun to melt away from the beach. I was thrilled to see a herd of musk oxen on the tundra, several jaegers in the air, and walrus on a beach.

Walt landed neatly on the runway of the abandoned DEW station.

We had begun unloading the plane when I made a discovery. "Oh, no!" I cried out. "I left *Flora of Alaska* in your library!" I was devastated, but it couldn't be helped. I'd just have to get along without that precious key. But, damn. It was my own fault, too.

Walt closed the hatch of his plane and waved, took to the air, then became a diminishing silhouette, a speck, and then nothing at all. We would be here for just over a week—no phone, no radio, no way of communicating at all.

Seals popped up on the beach, the vast Porcupine Herd (caribou) began drifting through, and an arctic fox came to smell my feet while I was washing my hair in a near-freezing puddle in the tundra. Flowers starred the grasses and sedge, and I knew most of them. But how I wished for that key!

The next day, I had just cleaned the kitchen after lunch (no countertop, no sink, no tables or chairs, no water, no heat, our own stove; no doors—the air itself was the fridge) and was heading out to look at flowers, when I heard a buzzing noise.

I looked up. In the sky to the west was a plane, a small plane. Astonished, I watched it land on the old runway only feet from where I stood. It was Walt, of course. The plane rolled to a stop and the pilot's side window slid open.

A hand and arm reached out. In that hand was a heavy book. Walt had brought my key! "Here you go, Dana!" he said. I grabbed the book, conveying my thanks. The window slid shut, the plane taxied briefly, and Walt was off.

Unflappable

I had always wanted to experience the high arctic and its 'round-the-clock daylight. In the late evening, the sun didn't set—it dipped near the horizon, held there for the length of two sunsets, and began to rise again. "Sundip," we called it.

Here above the Arctic Circle, I saw many of the animals I had hoped to see in the wild: polar bear, walrus, musk oxen, caribou, wolves, bearded seals, arctic fox—not to mention the fascinating birds: all three species of ptarmigan, skua, both kinds of jaegers, great snowy owls, longspurs, and eiders. Plants on the tundra were flowering like mad in the short summer season, and kept me occupied for days: tiny poppies, pink primroses, inch-high willows, cottontop sedges. There is nothing more exciting to a biologist than exploring a new ecosystem (unless it's discovering a new species).

One of the students, Joel, was an enthusiastic photographer, and he had brought state-of-the-art camera equipment, snapping pictures round the clock.

One evening as all of us returned from a hike up the Kongakut River, a mini lagoon off Beaufort Lagoon itself gave Joel an idea. Sundip was happening. Tired, we cooked dinner on our Coleman stove and made a pot of cocoa so we could take warm mugs outside, where we sat on the beach and looked out at the sea and the floating ice. The haunting cries of oldsquaw ducks and arctic loon came to us from the water, and thin golden clouds stacked themselves across the sun.

Another student, Marion, sat nearby on a driftwood log and began to play her flute. I remember the date—7/7/77—and the song she played, a song by Simon and Garfunkel, "Old Friends."

The clear, rambling melody floated far over the water and seemed to fall into rhythm with the wavelets, very small waves, because farther out to sea, the icepack still reigned.

Joel, sitting on the beach gravel with his mug of warm cocoa, looked east toward the lagoon, and had an idea. "See those swans in the lagoon?" he asked, pointing.

I looked. "They're tundra swans," I said, probably unnecessarily. "It's almost eleven p.m. Look, they're sleeping. They all have their heads tucked under a wing." Six swans slept peacefully in the center of the miniature lagoon on water like golden glass.

Joel had hopes of becoming a professional wildlife photographer, hopes rekindled by recent conversations he'd had with the two professional photographers from *National Geographic Magazine* who had been fog-marooned with us on Barter Island for several days during the previous week.

"I explored the shores of the lagoon yesterday morning," Joel said. "Until you get right out into the middle where the swans are, it's shallow. It's only two or three feet deep. It's not much deeper than that in the middle, maybe about six or eight feet. Now, think of it!" he said, his eyes never straying from the sleeping swans. "Tomorrow night, I wait until the swans are asleep. Then I take my camera and wade out toward them. With any luck, I can get pretty close before they spook, close enough for the telephoto, anyway. When the swans notice me, they'll get up and fly. I'll be close, and when they take off from the water—think what a fantastic photo that will be!" A dreamy smile lit his face. "I'll put the camera on automatic shutter, and get a whole bunch of pictures, bam, bam, bam, bam. Award-winning photographer, that will be me!"

"Sounds good," I said. "Go for it!"

The next day went by serenely. I keyed out wildflowers and paced a band of caribou moving slowly to the west. On the beach, I watched a bearded seal haul itself out of the water onto an ice floe. A flock of king eiders settled into the water of the mini lagoon but didn't stay long.

Dinner that night was beef stew, and we wolfed it down as if we were starving; barely-above-freezing temperatures make people hungry. Taking its time, the sun slid toward the horizon.

Eventually, Joel picked up his camera and told us, "I'm off."

The rest of us took our cocoa out to the beach and sat where we could watch his progress. The swans had returned to the lagoon, sleeping peacefully in the center just as they had the night before. Joel walked to the shore, took off his boots and socks, and rolled up his jeans.

With the camera slung around his neck, he began to wade toward the swans. Joel moved slowly, causing hardly a ripple. He drew closer and closer to the swans. They continued to sleep.

"I can't believe he's so close," Marion said. "I bet he gets some great pictures!"

"I hope it doesn't take him much longer," I said. "His feet will go numb. That water is close to freezing."

Joel moved closer and closer to the swans.

"He'd better not go any closer," Marion said, shading her eyes with a hand. "If he gets any closer, the swans won't fit into the frame."

Joel was now about twenty feet away from the swans. With one hand, he lifted the camera to his eye, the other hand he waved wildly. "Yahoo!" he shouted at the top of his lungs. "Yahoo! Yippee! Hi! Hi!"

The swans didn't move.

Joel began to slap the water with his free hand. He continued to shout, only it was now a scream. "Eeeee, you buggers!" He stood, thigh-deep, less than fifteen feet from the swans. They didn't even take their heads from under their wings.

After ten minutes of shouting, Joel slogged back to the shore, defeated. We saw him flop down onto the tundra to put on his socks and boots. Back at camp, he plodded past us into the old DEW Line building. "I don't get it," he muttered. "Going to bed."

Marion reached for her flute. "Lullaby for the swans," she said, and began to play "Old Friends" again.

Whenever I hear this song, I remember Marion sitting on a sea-bleached log, a log that must have come to the sea from somewhere in the interior of Canada, floating down the Mackenzie River, hundreds of miles from the treeless tundra. Faint cries of oldsquaw and arctic loon seem to punctuate the music as I hear it in my mind once again. "Old Friends."

Sealed

Midwinter arrived, and it was hot.

Eric, my then-husband, and I were on the last field expedition that we took from Millsaps College, driving from Mississippi into the wilds of the Sonoran Desert in western Mexico, bouncing along in the Biology Department's van, having camped our way across the South. We were four-and-a-half-hours' journey south of the Mexico border at this point, on a small, one-vehicle, two-week trip with six students. Desert ecology was the theme.

Having lived in Tucson for five years before moving to Mississippi, we knew and loved the Sonoran Desert, with its tall cacti, vast creosote-bush flats, unique bird life, and occasional flushes of colorful flowers.

This time, we had camped near the beach a mile or so from Bahia San Carlos (*bahia* means "bay"), near Guaymas, a city in Mexico, just north of the distinctive butte called *Tetas de Cabras*. In the mid-70s, there were no homes on this beachfront flat, only a strange network of sandy, semi-disappearing semi-roads, remnants of the set of the movie *Catch-22*, much of which had been filmed here several years before.

The partly rocky, partly sandy beach seemed to stretch away forever. We were alone, not another soul to be seen in any direction.

That first morning, I wandered up into the hills to refresh my memory, re-learning the vegetation I would teach to the students later, busying myself while they and my then-husband snorkeled in the pristine waters off the beach.

One of my favorite trees grows here, *Acacia willardiana*, mundane name for a most graceful and striking tree. Imagine a very tall tree, slender trunk, with stark-white bark and downflowing branches, rising from a hillside of cacti and small shrubs like a misplaced angel. That's *Acacia willardiana*. Another favorite is *Bursera microphylla*, a squatty tree sometimes called elephant tree, with tiny leaves and waxy, aromatic bark, in the same plant family as the Biblical tree/bush that produces myrrh. And a third favorite, *Fouquieria macdougalii*, appears as a cluster of narrow and spiny wands that flower scarlet at the tips every spring, a beloved of hummingbirds. F. macdougalii looks like the ordinary ocotillo of southern Arizona—but on steroids. Last, but far from least, there's the giant multiple-columned cactus, the cardon, hugely fat and incongruous in this land of sparse resources. I suppose that many people would drive casually through this arid country, bored by it all, but to me it's another kingdom, inhabited by exotic birds like the beautiful hooded oriole, and creatures like the regal horned lizard and the white-tailed antelope squirrel.

Tetas de Cabras Butte near Guaymas,
Sonora, Mexico

After several hours, the heat got to me, and I returned to camp for water. "Odd," I said aloud. "The guys—" (This trip happened to be all male, except for me) "—aren't back from snorkeling yet. Well, they'll be tired, too. I'll make lunch." I set up the folding table, made a huge pile of sandwiches, put out several bunches of bananas, and covered it all with the anti-fly netting we always carried.

I waited. And waited.

The sandwiches were drying. I became irritated. In some odd way, I had got butter smeared on my glasses while making the sandwiches and couldn't find the dish soap to clean them off. By now, I was getting a little miffed.

Leaving my glasses on the table, I marched down to the beach and stood on the rocky bank where I had seen them dive in. Did the guys think this expedition was all play? Two of them were on the lunch committee, and I had just done their work. *I'll give them a piece of my mind*, I thought as I looked out over the water. I should have worn my glasses, but even without them, I saw only blue water. Nothing.

Then I saw movement. Seven wet heads popped up just below me. "Hey!" I said sharply. "Let's have lunch before the sandwiches are ruined."

To my astonishment, the seven of them turned over and, showing me a row of wet behinds, dove and swam away. "Come back!" I shouted, irritation edging into fury. How dare they!

"Dana." From behind me came the calm voice of David, one of the students.

I turned and saw, blurrily, all six of the other fellows. "But—but—" I stammered.

"Dana," David went on. "You were talking to seals."

They never let me live that down.

To this day, I don't go ANYWHERE, not even to the bathroom in my own home—without my glasses.

A Night in the Honeymoon Suite

Part One

It's 1981. Picture this for me if you will: me at 40, a male college student, and a honeymoon suite. Yes, it happened. Yes, it was weird. In all the college biology expeditions during my life, I think this was the only time an expedition car caravan truly failed.

The professor who had planned and was supposed to lead a College of Idaho field semester to Australia had found another job, and with increasing concern, the college was searching for another professor to fill the position at nearly the last minute—someone who knew field biology and who could be trusted to lead an expedition in another country, teach on the road, and bring the students, who would be carrying a full course load in biology, back alive. Eric was hired, and left his current job to return, after more than a decade, to the College of Idaho. Having taken students on many expeditions, we were confident that we could pull this off, though we had never been to Australia.

Besides the two of us, there were about two dozen students signed up for the expedition, plus two teaching assistants. Both teaching assistants had been on a College of Idaho field expedition to Australia once before, as students, and that added to our confidence.

And on the last day of August—late winter in Australia—we were off!

After a few days of orientation in Sydney, we hit the road with two rented vehicles, a 22-passenger Toyota sightseeing bus and a little Toyota station wagon. The bus pulled an equipment trailer, and the station wagon sported a roof rack that held some of our duffel bags, covered with a bright blue tarp that I'd found in a Sydney sporting goods store. We needed to carry food, dishes, pots and pans, microscopes, books, calculators, binoculars, personal clothing and toiletries, dive gear, livetraps, sleeping bags, camp stoves, camp tables, stools, and tents.

After leaving Sydney, we spent a few nights camping along the coast, and began to drive north on the eastern coastal highway. We planned a longer stay in the Gibraltar Mountains of northern New South Wales.

We had more than 150 miles to drive before turning inland and climbing into the Gibraltar Range—and once there, we'd find no grocery stores. Therefore, in midmorning we stopped at a rest area outside a medium-sized town and planned the day's strategy.

We were not well-accustomed to driving on the left, particularly with the bus and its trailer, and had decided that it would make more sense to have the bus stay at the rest area—and to send one of the teaching assistants (Craig, who had experience driving on the left), and me into town with the more maneuverable station

wagon to get groceries. Craig and I would fill the station wagon with food for our week in the Gibraltar Range. Then, the two of us would zoom out of the town, driving back to the rest area and the others. We'd meet the bus, load most of the food into the trailer, and continue caravanning north to the mountains.

Craig and I bought the food: We bought fresh and canned fruit, bread, pasta, crackers, meat, canned tuna and ham, loads of vegetables, both fresh and canned (we would have no ice in the mountains), jelly and jam, eggs, pancake mix, powdered milk and powdered potatoes, pudding mix, cookies, tea, coffee, cocoa—everything you can think of to feed twenty-plus active, hungry people for a week. The station wagon was riding low when we left town, but we had bought everything we thought we'd need.

Back to the rest area we drove.

And found no one there.

The bus and its trailer had vanished from the rest area. Craig and I scratched our heads. We hadn't taken that long buying the food. In fact, we had returned half an hour before the estimated time we had given those waiting with our bus.

We waited, and waited, and waited. "Maybe something happened, and they went to look for us," Craig postulated reasonably. "When they can't find us in town, they will eventually come back here." That sounded good.

Two hours crawled by. The sky clouded over, and a keen wind flapped the blue tarp over the duffels on our roof rack. I kept glancing at Craig. He looked flushed, and he had begun to cough a little. He was coming down with something, and it wasn't good. He rested his forehead on the steering wheel and dozed.

I wracked my brain. I didn't know anyone in New South Wales. I had no phone numbers for anyone in Australia (and, of course, this was pre-cellphone). I had about forty dollars in Australian currency in my pocket, money left over from the food buy, and I had my VISA card (not nearly as easy to use overseas in 1981 as it is now). At least I had my purse and my passport.

Finally, I woke Craig. "They aren't coming back," I said. "Something has happened. I think we should go to the next town and stop at the police station. It's not far. They can check the hospitals and would know about any car accidents that might have happened."

Craig coughed. "OK," he said, starting the Toyota. "The police. That's a good idea."

Luckily, we had just filled the station wagon with gas. Craig pulled out onto the highway, and we joined the stream of traffic going north. We noticed that many, many more cars were going north than were going south.

A few minutes later, in light rain, we pulled off the freeway and asked directions to the nearest police station.

The police were kind and interested as we explained our situation. With a few phone calls, they checked hospitals for us. No Yanks. They checked New South Wales traffic accidents for us. No Toyota buses. They told us they would pass the information on up the line to the other police stations in towns along our path to the Gibraltar Mountains. If we stopped at each police station, we could learn whether or not any new information had been gained. And if the other expeditioners contacted the police, then the police could bring us together. This sounded as good as it was going to get.

Craig and I took off once more, driving north. We drove for hours in slow, heavy traffic, and the day grew old. Clouds parted, the sun returned, and the shadows of eucalyptus trees across the highway grew long.

Craig was getting sicker. Alternately, he would shiver with chills and then flush with fever—and even so, even as he coughed, he kept assuring me that he was fine, fine. But I knew he was not.

We filled our gas tank once more, bringing our cash reserve even more dangerously low than it had already been. And we drove north on the crowded highway.

The sun was sinking in the west when we reached Coff's Harbour. Although we had a car stuffed with food, we had no stove, and I felt that Craig needed a hot meal, so we pulled into town.

Coff's Harbour is a beautiful seaside resort town, complete with a picturesque bay and many fine hotels. First, we stopped at the police station, hoping rather desperately that the other expeditioners had contacted the police. But no.

We had learned that restaurants closed early in Australia, and it was "eat now or not at all," so we had a simple, hot meal in an Indonesian restaurant. The food took a big bite from the last of our cash.

I asked the waitress about nearby hotels and was hoping I could find one that would take my VISA card. Craig was shivering and coughing over the last part of his dinner.

"Don't you realize that this is school break time?" The waitress stared at me, astonished. "There isn't a hotel room to be had on the entire coast. Most of the rooms have been booked for months."

My heart sank. Would we have to spend the night in the car? In early September it was cold. But presently, the waitress returned with the restaurant owner, a gaunt, tall woman who was wiping her hands on her long white apron.

"So, you need a room for the night and all you have is an American VISA card?" she asked me.

I nodded, miserable.

She looked at Craig with some intensity. "He doesn't look right," she commented.

"I know," I said. "Please. I need to get him into bed in a warm place." I explained to her about the college expedition and the missing bus with my husband and students.

The woman gave a sharp nod. "Happens that my mother runs a hotel a few blocks down the street," she said. "Half a tick, and I'll give her a call." She disappeared into the back of the restaurant.

When the proprietress returned, her smile stretched from ear to ear. "The luck of the Yanks," she commented. "Mum said her booking for the honeymoon suite cancelled not an hour ago."

"I imagine that's really expensive," I stammered. There wasn't much credit room left on that VISA card.

"No worries," she said. "At least, not tonight. Alice," she said to the waitress. "Go ahead and close up the restaurant. Come on," she said to us, "let's go. We'll walk, and you can come back later for your car—or you can just leave it here tonight. It will be safe. I'll introduce you to me mum. You'll like her."

We set off in the deep dusk, Craig coughing in the chilly air of a late winter evening near the water. In no time, we found ourselves in a lovely cream-washed lobby, lush with tropical plants. The place looked expensive.

The restaurant owner's Mum came around the check-in counter to greet us. Unlike her daughter, the mum was short, round, and pink-cheeked. "Thank you, dear," she said, looking up at her daughter from behind narrow, pinched glasses. "I'll take it from here." I liked her at once.

I dug in my purse for my VISA card, hoping it would be accepted.

"No, no," the mum told me, waving my purse away. "Not tonight. We'll worry about all that in the morning. Come on." She rummaged for a moment behind the counter and found a flashlight.

She led us back outdoors, around the side of a building, and up a flight of open stairs. It was dark now, very dark. The air was damp and smelled of the sea. A key rattled in a lock, and she swung open a heavy wooden door.

I gasped. The room was beautiful, exquisite even, painted in white and palest yellow. It wasn't a room, it was a suite, with a bright kitchen, gleaming bathroom, lavish bouquet of fresh flowers on a satiny wooden table, and a huge king-sized bed. In my dirty jeans and travel-creased sweatshirt, I felt like a thug.

Mum settled Craig into an overstuffed armchair printed with golden flowers and soft green leaves. "There's tea and milk," she told me. "Better get a cuppa down that boy right away. And there's cream and berries, eggs and bacon, pastries and butter, and fruit there in the box too—help yourselves."

By this time, I was close to tears. I felt a soft hand patting me on the shoulder. "She'll be right," Mum said. "I'll leave you now. Phone is on the wall. Call the number on the dial if you need me."

"How can I thank you?" I began, but Mum was closing the door, already gone. I stared out the picture window for a moment and gasped again. Lights blinked far across a mirror of dark water. We were on the bay—*right* on the bay.

I made tea, laced it with milk and sugar, and handed a cup

to Craig before taking one for myself. He was still shivering, and his cough was worse. "Tea, hot shower, and off to bed for you," I told him. I'd found two aspirin in the bottom of my purse and held them out to him

"But Dana," he said, protesting. "I'm going to sleep on the floor. I mean, we can't share a bed. We just can't."

"You are not sleeping on the floor," I said, too tired to argue, "and neither am I. I'll figure out a way to fold the covers so we don't touch. Besides, I have no intention of getting close enough to catch your cough."

Craig headed to the bathroom. I hoped that the steam of the shower would help clear whatever respiratory bug he'd caught.

Sitting at the table near the window, I cradled my teacup in both hands and gazed out across the water. At least, through the kindness of strangers and a shining gem of luck, we had secured a safe place for the night, a place where at least Craig might not get worse.

I examined the floral centerpiece on the table, from which rose an unfamiliar fragrance. I was teaching a course called Flora of Eastern Australia, so I had studied Australian plants rather frantically during the three weeks we'd had after Eric had been hired by the College of Idaho and before the expedition had lifted off for Sydney. Banksia, I knew, and the full crimson heads of waratah. Fuzzy orange kangaroo paws shared the vase with sprigs of something white that I didn't recognize, and there were glossy fronds of bronze ferns. The flowers were beautiful, but they were alien.

What was I doing? What was I going to do tomorrow? I dumped the contents of my purse onto the polished tabletop and counted the money. Two dollars and forty-nine cents, a passport and driver's license, my lucky glass frog, and a VISA card for an American bank. That was it. Our flight home was scheduled for a date in the middle of December, and I didn't even have the tickets with me. A little over three and a half months in Australia, a VISA card with a couple of hundred dollars left that were available, and two dollars and forty-nine cents in cash.

We still had nearly a hundred miles to go before reaching the top of the Gibraltar Range where the campgrounds began, and somehow, we'd have to get gas for the station wagon at least once more.

And what if we made it to the mountains, these mountains neither of us had ever seen? The Gibraltar Range was not a pinpoint on a map. The mountains rambled over hundreds of square miles—forest and stream, canyon and ridge.

My lucky glass frog.

Followed by a small cloud of steam, Craig tottered out of the bathroom, wearing the same shorts and t-shirt he had worn going in. He made it to the bed, where I had pulled back the covers, and he fell onto the mattress like a zombie. I pulled the covers over him and turned out the light.

As I slipped off my shoes and climbed into the other side of the bed, I thought for the thousandth time, *What possessed the others to leave us like that? Why didn't they come back? Why didn't they go to the police? Why?* And I thought some very dark thoughts about my husband and the students and what I would do to them when we found them. *If.*

Over dinner, Craig and I had discussed a plan, and that plan was simply for us to get up into the Gibraltar Range, find a campground, and stay until the coming weekend was over, searching all the campgrounds. We had food—lots of food—and matches. I could cook over a fire in the food cans. If we didn't find the others, then we'd drive down out of the mountains and do something—*something*.

I tried to formulate a plan for the worst possible scenario. What if we never found them? Had they been kidnapped? Had the bus fallen over a cliff into the sea or tumbled down the side of a mountain into a hidden canyon? My thoughts got weirder as I glided away from consciousness. Had the others decided that they hated Craig and me and sent us after food so they could get away? How could I get money wired to me so we could go home? Failing that, what about a robbery? I didn't know the first thing about robberies. Or maybe I could stay with the mum and work in the hotel. But what on earth would I tell all those parents if their children had disappeared forever? Exhausted, I drifted off to sleep.

Part Two

Light from the rising sun crept across the bay and filled the pale rooms. Craig was still asleep, snoring. I took my shower and made a pot of tea. Then I heard a light tapping on the door. When I opened it, no one was there, but on the top step I saw a cardboard box. It felt warm in my hands. Inside rested pastries fresh from someone's oven. It was past time for that hot cup of tea.

The bay stretched for miles and to my delight, it was full of birds—stilts, ducks, sandpipers, gulls, swans—all unfamiliar: grey teal, black swan, Pacific black duck, musk duck, pink-eared duck, and many more—all species I had never seen before. *If nothing else,* I thought grimly, *I'd have a better bird list than those runaway jerks!*

As I sipped my tea, I was wondering how it would be possible for me to choke twenty people to death for abandoning us. Could I sneak up on them in the night and do it one at a time? That seemed the best option.

Craig woke up just as I finished making the bacon and eggs, and we had a feast as we sat at the gleaming wooden table and breakfasted on berries with cream, mangoes, pineapple, bacon and eggs, and hot pastries with butter.

He seemed somewhat better. And it was a gloriously sunny day.

I stuffed the two dollars and forty-nine cents, passport, lucky frog, driver's license, and VISA card back into my purse. We took some of the fruit and pastries with us.

The Mum was at the front desk and after a phone call she made, her bank agreed to honor my VISA card. She charged me a quarter of the going rate for the suite. I was so relieved that I was shaking.

After thanking the mum and checking in with the police one more time, we left Coff's Harbour.

We got onto the highway and headed north once more. While Craig drove, I scanned the southbound lanes for the Toyota bus and its little trailer. Where were those evil, evil buggers who had forgotten us without so much as a check-in with the police or contacting a hospital?

The car, filled with heavy food and topped with bulging duffel bags under the blue tarp, was so sluggish that our top speed was well below the speed limit. This was not going to be a fast ride.

After an in-car lunch of the fruit and pastries we had brought from the Coff's Harbour hotel, at last we made it to the final town before the road would begin to snake its way up into the mountains: Gosford.

Trying to radiate a confidence I didn't feel, I went into a Gosford bank and eventually persuaded an elderly clerk who was wearing suit coat, tie—and shorts—to give me fifty dollars cash from my VISA card. That accomplished, we filled the Toyota's tank with gas, checked in at the police station (no news), and soon found ourselves climbing up, and up, and up into mountains blue with a faint haze of eucalyptus smoke.

Craig switched on the radio. Unfamiliar pop music filled the station wagon. The car slowed as we took a series of tight turns, and slowed further as the road became steeper and steeper.

I rolled down my window and took in the fresh, astringent scent. These weren't my mountains. This wasn't my forest. Everything here seemed otherworldly, even the ladybug that landed on my arm, a ladybug that had too few spots and whose elytra were yellow rather than red or orange. But nevertheless— these were mountains, this was forest, and we were leaving the crowded coastal plain with its towns and traffic, far behind. We hadn't seen another car for miles. My spirits lifted just a little.

Craig negotiated a tight turn onto a bridge over a deep ravine. A song wafted from the radio and seemed to spread far over the patchwork valley below. I remember it to this day, and later I was able to track it down and buy the album. The song was "Every Woman in the World," by Air Supply.

I looked down on a feathered brown back and long wings soaring over the farmland below, my first sighting of a wedge-tailed eagle.

Suddenly the station wagon lurched to the left. Craig fought the wheel as I tried not to look over the 200-foot drop-off on our side of the road. *What now?* I thought in a panic. *Is the car breaking down? What will we do?*

Craig was more pragmatic. He brought the car to a stop on the road shoulder.

We got out and had a look. "Just as I thought," Craig said in disgust. "Blowout." Indeed, the front tire on the mountain drop-off side was not only flat, it was shredded.

Craig shrugged and started searching for the jack. He finally found it in the trunk under a case of green beans.

He bore down as hard as he could on the lever, and I joined him, but it wouldn't lift the car. It kept slipping. "Got to empty out all this food," he said. "Then we've got a better chance of getting the jack to work."

Below us, the wedge-tailed eagle soared on flat wings, far over the patchwork valley.

A strange tinkling sound fell on us like rain, like a shower of chimes from hundreds of tiny bells. I looked back toward the ravine we had just crossed and saw a huge, gnarled eucalyptus tree rising from the depths. Tiny birds, dozens of them, hopped among the branches. It took me a moment to realize that those little birds were singing the sound of bells.

I grabbed my bird guide. "Bell miners," I told Craig. "Those are bell miners!" I had hoped to see bell miners, but not like this!

Craig and I set to unloading the car on the road shoulder—bags of potatoes and carrots, cases of tuna and ham, jugs of punch-making syrup, cartons of powdered potatoes and milk, boxes of oranges and grapefruit.

"This is the last straw," I told him over my shoulder, heaving a box of canned corn from the depths of the trunk and setting it down in the gravel. "If we get to a campground up on top, I'm going to build a big fire, make some cocoa, and eat every cookie we bought. And if we find the others, I swear I'm going to kill them all."

Suddenly I noticed that Craig had stopped unloading and was standing still, stock-still, as if he had been flash-frozen. "What?" I said, irritated. "What?"

And then I saw it: a small Toyota bus gliding down and down the turns above us on the mountain. Someone sitting in the passenger seat suddenly rolled the window down and waved with both arms.

"It's them!" I said unnecessarily, and sat down in the dust as if I had been punched in the stomach.

It definitely felt like I had been. To be exact, it was Eric driving, and one of the students, Brad, waving from the passenger seat.

Our 1981 camp in a eucalyptus grove, Gibraltar Mountains, New South Wales, Australia.

Craig and I sat dumbly and watched as the two of them jacked up the station wagon, changed the tire, and loaded some of the food into the bus. They turned the bus around and took off. In silence, Craig and I drove up the mountain after them.

After less than ten miles we pulled into a campground, a lovely, clean space of leaf litter in the dappled shade of tall eucalyptus. There were our tents. There were our two camp stoves and camp tables. And there were the students.

They crowded around as Craig and I got ourselves out of the car. "What happened?" the students asked. "Why did you leave us at that rest stop?"

My blood pressure boiled over. "Why did we leave YOU?" I almost screamed. "We didn't leave you. We bought the food, got gas, and went back to the rest stop half an hour earlier than we had hoped—and you guys were gone. Why did YOU leave US?"

The students looked at one another in dismay. For a moment, no one spoke. Then the other teaching assistant, Dave, ventured, "We'd been at the rest stop for about twenty minutes. Then we saw a gray Toyota station wagon with a blue tarp on top go flying past the rest stop, going north."

"It was you," a student put in.

"So, we jumped into the bus and followed you," Dave went on. "We followed you for about forty miles before we lost you. The bus couldn't keep up. You were going like a bat out of hell." There were nods all around.

"And we kept going and going, and it was raining in all that slow traffic and getting dark, and we were tired," Eric chimed in. "So, we got some fast food and stopped at a caravan park for the night." He named a little town north of Coff's Harbour.

"It was miserable there," one of the students said earnestly. "Cold and wet and windy. In the morning we took off to get here. We set up camp and then the Doc and Brad got in the bus and went down to find you."

I did some heavy breathing and some internal *grrrs*. "It never occurred to you to stop and ask about us at a police station?" I couldn't help saying. "Or to have the police check for accidents and in hospitals?"

Heads were shaken. "Noooo," someone said.

"How could you think we could buy a week's worth of food in a strange town in twenty minutes?" I had to ask. After all, this expedition was supposed to be a learning experience. "Or that we would take off past you without stopping?"

"Well, we found you, and that's all that matters," Eric said. "And besides, there was that blue tarp on the roof. I mean, nobody has a blue tarp. That's the only blue tarp I've ever seen."

Hmmm, I thought. I had never seen a blue tarp before arriving in Australia, true (though now they are common in the USA). But Craig and I had seen blue tarps everywhere as we had driven north—atop haystacks, on trailers, used as awnings, at construction sites—all over New South Wales.

"You must have spent a terribly cold and uncomfortable night in the car," one of the girls said. "All the sleeping bags were with us. And I bet the car was so full of food that you couldn't even put the seats back."

"Actually," I told her, "we spent last night at a luxury resort hotel."

The Voice of Kath Walker

On my first Australia expedition, I took personal books in addition to field guides and books in our traveling biology library for the students. I must have books, and back then, of course, there was no such thing as an ebook reader. So, I took these books—Tolkien's *The Lord of the Rings*, *The Book of Morgaine* by C.J. Cherryh, and *My People*, by Kath Walker.

You may know the first two books, but let me tell you a little about Kath Walker and *My People*. Kath was born in 1920 and became a maid in a Brisbane household. Of aboriginal tribal origin, she did not know how to read. Intrigued by the books she saw in her employers' home, she tried to learn. And the mistress of the house saw that burning desire and taught her how to read. In 1988, Kath took back her native name, and is now known as Oodgeroo Noonuccal. She died in 1993.

I do not need to go into the history of aboriginal peoples in Australia, but many, like Kath, had no voice and little hope. Kath became that voice and that hope, and her award-winning poetry spoke of her people and for her people. She reached a wider audience as well, people like me, people in love with words and the natural world. So, when I went to Australia for the first time, I took her book with us, the slim volume of poems titled *My People*, intending to reread it while camping in her native country, also intending to share the poems with my students.

This Australia trip had been planned by someone else, and he had a different idea of how to do a field expedition. He had scheduled, and paid for, a week's rental of an old miners' barracks on a large island just off the east coast near Brisbane, a twenty-minute ferry ride from the edge of the city. This barracks had tiny, hotel-like rooms and a large kitchen/dining hall. We preferred to tent-camp in the wild as much as possible, but the lodging fee on Stradbroke Island had been pre-paid, so off we went.

Eagerly, I leaned on the railing near the bow of the ferry, hoping to see birds new to me on the short voyage to the island. I wore my small daypack, crammed with necessities such as Chapstick, purse, bandana, *A Field Guide to the Birds of Australia* by Graham Pizzey, sunblock, and a sweatshirt. Also in the daypack was my copy of *My People*.

After we left the shore and reached the middle of the sea passage, birds became scarce, and I retreated to a deck chair and watched the people.

An old woman caught my eye. She looks like an aboriginal, I thought. I had never met an aboriginal. As if deep in thought, the woman leaned over the rail, her refined profile shadowed by a wide-brimmed hat. She looked familiar. I couldn't make out why I felt that I had met her before. Then, with a sudden revelation, I dug into my daypack and pulled out the copy of *My People*, turning to the back cover where there was a small photo of the author.

I compared the photo to the woman standing at the rail. Could she be Kath Walker? I knew from the blurb in my book that Kath had been born on Stradbroke Island. Could it be?

Clenching my jaws, I approached her, holding the book. "Are— could you be Kath Walker?" I stammered.

"I am," she said, turning to me with a smile.

"Could you please sign my book?"

She did.

Soon we were sitting side-by-side in deck chairs, and she was pummeling me with questions. Where were we from? How many of us were there? What would the students be doing? Why were we coming to Stradbroke? Where would we be staying on the island? How long would we be there?

And suddenly, I was given an invitation. "When Australia declared me to be a national treasure," Kath said with a wink, "they gave me a few acres on Straddie, and this is what I do: I bring over inner-city aboriginal children, and they stay in one of my little caravans (trailers) for a while. I teach them the ways of our people—the myths and legends, the wild foods, how to track animals and build fires—all those things that were forgotten when they moved to the cities. You must come," she insisted. "Everybody on Straddie knows how to get to my place. Moongalba, it's called. Come in three days, just after noon, for a visit, just you."

I couldn't believe my luck.

We drove our two vehicles off the ferry, found the miners' barracks, and unloaded our gear. While most of the students settled into their rooms, I took three of them and our station wagon, and we went grocery shopping. The clerk at the store gave me directions to Kath Walker's place, her "compound" he called it.

We spent two days roaming the island, birdwatching, tidepooling (the deadly blue-ringed octopus is beautiful, but it is startling when a student hands you the creature and asks you what it is), learning the trees, and cooking our meals in the large, well-equipped kitchen/cafeteria. We learned much, though we felt that we would be happy to get on the road again soon, away from so many people, traffic, and homes.

But on the third day, I had one of the graduate assistants drop me off on the dirt path that led to Moongalba.

I found Kath holding court outside a small silver caravan trailer hidden in a grove of eucalyptus trees. On a wooden table sat a basket tray, and on the tray, inching around on some flat leaves, were several large beetle larvae. Fat, pale, and caterpillar-like, each one was about six inches long, segmented, and puffy, with a sharply dark head. "Are those witchetty grubs?" I asked.

"They are," said Kath from a nearby lawn chair. "Want to try one?"

Gingerly I picked up one of the grubs, which, perhaps knowing it was in danger, began writhing itself in and out of the shape of the letter C. I bit off the head, discarded it, and ate the rest of the body. The grub was slippery and starchy, with little flavor.

"I gathered these because I have a new batch of children arriving tomorrow on the ferry," she said. "We're going to eat these and learn where to find more."

Then I noticed the two other people, whose lawn chairs had

been placed in deep shade. "Meet Mr. Z and Ms. X (I do not remember their names)," Kath was saying. "They declined the treat." Indeed, Ms. X was staring at me wide-eyed, with one hand covering her mouth.

"Mr. Z and Ms. X are reporters from Sydney," Kath explained as she found a chair for me and pulled it into place. Both wore nice clothing, city clothing, whereas Kath and I had on well-worn jeans and t-shirts. "I'm just about to send the proofs of my new book back to the publisher," Kath continued. "I have the original illustrations here on the table, and Mr. Z and Ms. X have come to interview me about the book."

She reached for a stack of brightly colored drawings on very thick paper and handed one to Mr. Z. "What do you make of that?" she asked him.

"Well, the five-pointed star is important in many of the mythologies of the world," Mr. Z began. "It's a star in a blue sky." Then he stalled. He handed the drawing to Ms. X.

"There are roads inside the star," Ms. X contributed brightly. "Or maybe they are trails. They keep to the inside boundaries of the star. And outside is the blue of the sky."

Ms. X gave the drawing back to Kath, who passed it to me. "And what do you see?" she demanded.

I swallowed nervously, but when I saw the drawing, without thinking I said, "It's a starfish. We're looking right down into it, as if we have x-ray vision. That round thing is the madreporite, with the ring canal attached, and there's a radial canal going down each arm, with the little tube feet sticking out. The blue is the sea."

Kath threw back her head and laughed. "Got it," she said. The reporters fidgeted in their chairs. "Not to worry," she said to them, "just remember, symbolism comes from the real world."

"What will the book be called?" I asked.

"*Mother Earth and Father Sky*," Kath replied. "It will be out in a few weeks."

For an hour or so, Kath chatted about the other illustrations and about the things she would teach the city children after they arrived on the ferry the following day.

Eventually, I sensed that it was time to leave and rose; I had told my driver to come back in about an hour and didn't want him to sit in the hot sun waiting at the entrance to Moongalba for too long. As I said my thanks and goodbyes, on impulse I asked Kath if she would be willing to give a poetry reading for the students. "Of course," she said. "How about the day after tomorrow, after I have the new children settled in, at about dark? I know where you are."

Again, I couldn't believe my luck.

The two days passed quickly. One student captured an echidna, found an old aquarium in a storeroom, half-filled it with dirt, and deposited the creature inside, with a heavy board and a cinder block on top to keep him in. I told the student that he could keep the echidna overnight but then he would then have to let him go. The echidna, named Sundance (the Sundance e-KID-na, of course) dug into the dirt, ate the termites we found for him, and during the night threw off the board and cinderblock and clawed his way to freedom through a solid wooden door.

We found jellyfish, water lilies, and lavender janthina sea snails, and one afternoon had a swim in a pond we found in the forest. But at last, it was *the night* and Kath appeared at the cafeteria door in the deep twilight carrying a book and a large black case with a handle.

In the dusky cafeteria, she read a selection of poems from *My People*. I was enchanted. Her face was a shadow, scarcely to be seen in the dimly lit room, but her voice—her voice was everything.

And after the reading, she opened the black case and announced, "Now for the movies."

In the case were a projector and a reel of film.

"I have been making films with the help of Judith Wright," she told us. I explained to the students that Judith Wright was a noted poet of Australia, who, like Kath herself, was a strong advocate of aboriginal rights.

"We have been recording the old songs, myths, and dances of my people," Kath explained, "getting them on film before they pass out of memory. Here are the first few we have done. As yet, they are unedited, but they are the real thing."

We began watching.

A painted man appeared out of the darkness near a fire under a small tree. He danced, and while dancing, spoke in an unfamiliar tongue. He carried a short spear and gestured with it, stabbing at the air. Suddenly, he turned his back to the camera, and he was done. After an interval of blank darkness, a different painted man appeared, this one wearing the skinned-out head of a kangaroo. He, too, told some story in his native language, then turned his back and was done.

We watched a dozen such performances, powerful and mysterious.

I helped Kath pack up her equipment and walked her outside. She laid a finger to my cheek and said, "You wept. This is why I am a voice."

Refusing the offer of a ride, she walked alone down the dirt driveway in the starry darkness.

The Great Mince Disaster

When we boarded the cargo barge *Robert Poulson* to head to Heron Island for a month's stay, I knew we'd have to spend Thanksgiving on the island, but wasn't too worried. This stay on Heron was to be the finale of our first semester-long Australia field expedition. Just imagine how excited I was. I would be teaching an intensive short course titled Fishes of the Great Barrier Reef—such a dream (I have loved fish of all kinds since I was very small).

On that Halloween night in 1981, I hadn't been to Heron Island before, but I had been told that the barge *Robert Poulson* would be making the trip from Gladstone, a town on the mainland coast, to Heron, once a week. From the island, we could radio the ship for what groceries and other supplies we needed, and for a small fee, a ship's steward would shop for us and load them onto the ship.

In 1981, Gladstone was a small city, but I had shopped there before getting on the ship, so I had some idea of what I could order in the way of food.

We boarded the *Robert Poulson* in the early twilight on Halloween. After searching several markets, I had found one small pumpkin. I carved a face into it, of course. I put in a candle and wedged the pumpkin between some rigging and a corner of roofing just above the pilot's cabin door. With the chug-chug of the motor and the rocking of the waves, the pumpkin grew a long, white string of waxy drool during the night's passage.

The fifty-mile trip took all night, and we spread out our sleeping bags in the flat cargo area. In the morning, the captain woke us by playing calypso music. We looked out over a sparkling turquoise sea and there it was, Heron Island, our home for all of November. We passed through a channel in the reef, docked, and began unloading our supplies and equipment.

Heron is only forty acres in area, but the island has three distinct parts, nevertheless. One part is a small resort—not the place for people who wish to idle in the sun and show off fancy clothes, but a haven for divers and others whose desire is to explore the reef itself. Another part, dominated by native shrubs and pisonia trees, is government-designated wilderness. And the third, the middle part of the island, is a university research station with tiny tin cabins, a library, laboratories with piped-in seawater, and a doorless building that combines cooking, cafeteria, and classroom seating. The latter would be our headquarters. We moved in. It was paradise.

For a while.

On a sunny, windless day, I sat in a twelve-foot aluminum boat with two students, on the water above Wistari Reef, fifty miles from the Australian continent, fishing half a mile from Heron Island, fishing with some desperation and keeping almost every fish we caught—until Brad hooked a gray reef shark nearly as long as our rowboat. Brad was excited. So was the other student, Eddie. I, of course, was terrified. One sharp bump of the shark's nose on that aluminum boat would have dumped all of us into the sea with the shark. After a couple of tense minutes, Brad cut the shark loose. Shaken, we rowed back to Heron with the day's catch. But—this was a college ecology semester. So why were we out fishing?

We were fishing for food.

Our kitchen had several door openings to the outside but no doors in them, so sometimes there was a surprise, including a crab that came to dinner uninvited, or the black, seagull-sized shearwater that thumped me in the back as it tumbled into the kitchen while trying to land in the dark after being at sea all day (shearwaters are great at flying but awkward at landing).

Our supplies came once a week on the *Robert Poulson*. I'll not soon forget the *Robert Poulson*—sturdy, broad-beamed, and pristine white against the turquoise of the tropical sea, beautifully kept, slow but steady and reliable. I saw my first flying fish from the deck of the *Robert Poulson*, and my first manta rays and humpback whales. It took the *Robert Poulson* about eight hours to make the trip from coastal Gladstone to Heron Island.

College students have healthy appetites. Typically, appetites fall off in tropical heat. Before coming to Heron, we had been camping in the mainland tropics for weeks and had been accustomed to less-than-gargantuan appetites for some time.

But we soon discovered that college students who are diving or snorkeling a reef have *huge* appetites. We had brought as much food as we could with us, and soon were making giant-size tuna noodle casseroles, massive batches of very cheesy mac and cheese, mammoth trays of fruity Jell-O, washtubs full of fruit salad, lots of scrambled eggs, bacon, sausage, and pancakes—plus cakes, pies, hamburger goulash, hamburger shepherd's pie, spaghetti with hamburger sauce, hamburger mac and cheese, hamburger lasagna—you get the idea. We went through food at a rapid rate. And, of course, we used up our perishables swiftly.

After our first week on Heron, it was time to order a batch of food for the faithful *Robert Poulson* to bring us on its next visit. It worked like this: one of the Heron Island research staffers had a radio in his cabin. Using that radio, I'd call the crew of the *Robert Poulson* on the mainland, and for a small fee plus the cost of the food, one of the crew would go to a mom-and-pop grocery near the harbor at Gladstone and pick up items on the list that I would read to him over the radio. In 1981, Gladstone was not a big town, only about 20,000 people, and there were no large supermarkets.

After a week of hearty appetites, we had run out of hamburger, lettuce, fresh fruit, green beans, eggs, and so on. Tuesday morning, I sat in the kitchen and wrote up my list.

After a couple of months traveling in Australia while feeding these hungry students, I'd learned a few things—how to buy half a sheep at a butcher's and how to have him cut it up for me; that Weetabix was a good breakfast cereal; that cookies were called bikkies; what golden syrup was; what was meant by a Pavlova; that Australian hot dogs were strange and were like American hot dogs in shape *only*. And I had learned that there was no such thing as "hamburger," that is, ground beef. The term for ground beef was "minced beef," usually called just "mince." Also, the metric system was the form of measurement used, so groceries were sold not by the pound but by the kilo (a kilo is just over 2.2 pounds).

Armed with this knowledge, I completed my shopping list and made my way down the sandy path between giant pisonia trees to the tin cabin where the radio lived.

The research biologist there welcomed me. I had the impression that he was accustomed to such visits. He sat me at the radio. He showed me how to use the mic, and keyed in the call signature for the Gladstone office of the *Robert Poulson*. He handed me the mic.

What I heard was static. I finally reached one of the *Robert Poulson's* crew and began to read my list to him. That day, the air was seriously growly with static. Evidently the crewman was finding it just as difficult to hear what I was saying as I found trying to understand his replies. "Six pineapples, over," I would say—and then I'd have to repeat myself several times. "Ten onions, over. Got that, over? Ten onions, over. Got that, over? Ten onions, over. Got that, over? OK. Four watermelons, over." But the crewman was both kind and dogged, and in half an hour we got it done, right down to my last item, 20 kilos of mince.

"Might not be able to get all these quantities," the crewman said then, in a short interval without static. "It's just a little mom-and-popper store that we deal with."

"Please, just do what you can," I told him. "It's fine to improvise and substitute. And thank you so much." And we signed off.

I said my good-byes to the researcher and, feeling quite efficient and successful, made my way to my cabin, changed into a swimsuit, and spent a lovely afternoon learning to identify corals on the reef.

The next morning one of the students, Brad, came running to the kitchen from the beach. "The *Robert Poulson* is here," he shouted.

I gathered about a dozen students, and we made a human "box brigade" out through the lagoon to the *Robert Poulson*—at low tide, she had too much draft to make it over the reef and too much beam to use the channel to get to the dock. The crew handed us bags and boxes from the deck as we stood in the water, and we passed them, one by one, down the line of people and across the lagoon until the supplies made it to the beach. The crew waved good-bye, and as the *Robert Poulson* chugged west toward the horizon, the students picked up boxes and bags from the heap on the sand and toted them to our big kitchen.

Several students stayed to help put things away. Milk, fresh green beans, lettuce, cabbage, and butter went into the fridge. Pineapples and melons went onto the shelves. Noodles and potato flakes went into the cabinets. "Where's the hamburger?" Albert, one of the students, asked. We looked around the kitchen for my 20 kilos of mince.

Into a corner next to the stove, someone had slung a large—a very large—transparent plastic bag. Inside I could see what looked like several small bales or shaggy bunches of something green and leafy.

Puzzled, I stuck in an arm and pulled out a bunch of—of *mint*. "Mint?" I said to Albert, who was still looking for the 20 kilos of mince. I held out the bunch to show him. "I didn't order any mint."

Damply stuck to the bottom of the plastic bag was a square of white paper. I fished it out and found a note, written in pencil. "Sorry," it read. "I looked all over town, but I could find only four kilos of mint."

Mint? *Mint?* Suddenly I sat on the sandy linoleum floor and laughed until the tears ran down my cheeks. "Mince!" I said to Albert, who looked at me as if I were demented. "Mince! Mint! Static!"

I picked myself off the floor and swatted sand from the seat of my swimsuit. Of course. The radio. I had said "mince," and the crewman on the other end of the line had heard "mint."

We crammed the mint into the refrigerator.

But we had to have protein, and we had a seven-day hamburgerless wait before we could get more. We scrounged some battered fishing tackle and some bait from another researcher, and every day, borrowed the station's twelve-foot boat for two hours. Three of us would motor the half-mile over to the next reef, Wistari, which was not a sanctuary reef like Heron, and there we would fish our hearts out.

The first day out, Brad hooked that shark.

After we got used to fishing on Wistari Reef, we became quite successful and caught coral trout, red-and-white emperors, sea bass, grouper, various cods, toothy wrasse with blue bones, silver bream, and a number of other edible fish. During the week before the *Robert Poulson* came around again, we ate fish with mac and cheese, fish and pancakes, fish lasagna, fish goulash, fried fish, baked fish, fish hash, and other fish delectables.

And I might add—we drank mint water every day. Our Jell-O salads rested on platters bordered generously with fringes of mint. Fried fish and baked fish were served with mint sauce. Green beans were served with mint sauce. Canned corn was served with mint sauce. Every serving dish and every person's plate came from the kitchen garnished with sprigs of mint.

Sometimes I still dream about mint. Mint and the turquoise waves, the big reef shark, and the rowboat rocking and rocking over Wistari Reef. "Look, Brad, you've got a blue tuskfish on the line!"

A Yank Thanksgiving

'm not a diver, but I found that I didn't need to dive to explore the reef at Heron Island. Snorkeling brought me many wonders. Fish, corals, marine worms, shellfish, crabs, isopods, starfish, squid—we saw countless species and learned how to identify many.

We met the marine biologists and graduate students working at the station, too.

One night we witnessed the first predicted mass spawning of reef corals. Two coral scientists from New Guinea had studied all the previous times on record when the reef coral had spawned and had made calculations. They had used the phase of the moon, the season, the tide schedule, and other factors to determine which night they thought the spawning would take place.

Night fell early and swiftly, and at the predicted time on the predicted date, we joined the two scientists and their graduate student on the beach, all of us hoping.

The tide was out. Some distance out to sea, surf crashed on the exposed reef crest. Here and there in the shallows close to shore, tiny points of phosphorescence gleamed and blinked, over and over.

At low tide dead water, the lowest point of the tide, the sea calmed and then we smelled something new: a strong odor of fish, almost overwhelming.

With flashlights aimed at the bottom, we waded slowly into the water until we came to the first corals.

Clouds of white sperm billowed up from the corals—not from just one species but from *every* species. Small pinkish eggs floated up through the mist of sperm and rocked on the surface.

Plate corals spawned, staghorn corals spawned, pore corals spawned, brain corals spawned, mushroom corals spawned. There are hundreds of species of hard corals on Heron Reef, and this was their night. Everywhere, the tiny eggs floated up through the sperm-clouded water, until the surface was covered with them.

The tides would move them. Fish and invertebrates would feast. After a time, each remaining fertilized egg would hatch into a tiny jellyfish-like creature, which would float on the tides and swim weakly until it came to something solid. Then the little medusa would attach and would start building a skeleton.

No one spoke as we waded and watched, flashlight beams wavering into the clear water. This was a wonder of science, a triumph of study and logic. But it was also a triumph of the wild.

The sea is the world's last great wilderness. When you dip your toes into the sea, you stand on the edge of a vast mystery—fascinating, inspiring, frightening, and humbling.

After a space of several hours, we walked back to our little tin cabins in silence.

It's difficult to get into the swing of worldly concerns after such an experience, but Thanksgiving was coming, and soon.

Late the next afternoon I went to the radio shack and called the *Robert Poulson* with my shopping list: cranberry sauce, two big turkeys, many loaves of bread, cans of cinnamon, cloves, ginger, basil, and sage; butter, canned pumpkin, pie crust mix, cans of evaporated milk, eggs, sugar, green beans, cornstarch for making gravy, heavy cream, and then the ordinary foodstuffs we would eat on the other days. The *Robert Poulson* was to sail from Gladstone in two days and would be at Heron in three.

We'd make stuffing and roast the turkeys. We'd have pumpkin pie with whipped cream. Thanksgiving would be a holiday for the students: no lectures, no reef study, no tests.

The next day, a graduate student came with a message from the *Robert Poulson*: I was to call back immediately. There was a problem. I ran to the radio shack and called in.

Well, I was a Yank, that was the problem. As a Yank, I hadn't realized that Thanksgiving on the last Thursday in November is a Yank thing, not an Oz thing. Grocery stores in Gladstone were not filled with turkeys and cranberry sauce and canned pumpkin as I had unconsciously expected them to be. Instead, they offered their usual abundance of bananas, mangoes, and fish. I had a long conversation with the steward, and he promised to do the best he could.

When the *Robert Poulson* chugged into view the following day, I was filled with apprehension. The students unloaded our boxes of supplies and ferried them to the kitchen.

The steward had done admirably. Instead of canned cranberry sauce, he had supplied jars of orange marmalade and berry jam, suggesting that we mix them (this we did; the combination was delicious). There were no turkeys, but in one box we found fifteen whole chickens. Those would have to do. Pie crust mix, spices, eggs, bread, and sugar were there in abundance, and so were cans of evaporated milk and a large bag of green beans. Instead of canned pumpkin, the steward bought us several plump orange squashes. We'd have to gut and bake them, then pulp the flesh and strain it before we could make pies. That was do-able.

I reached out suddenly and leaned on the stove, the big four-burner stove with its nice, ordinary-sized oven. How was I going to roast fifteen stuffed chickens and bake ten to twelve pies on the same day in only one oven? I walked into the cafeteria section of the building and slumped into a chair. Then it hit me.

The coral scientists, other researchers, and a handful of graduate students were also staying at the research station. Their cabins had tiny kitchens, and those tiny kitchens had ordinary-sized stoves. Leaving the students to put away the groceries, I made the rounds.

"We'd like to invite you to our Thanksgiving dinner tomorrow afternoon," I said over and over. I then gave my best begging smile. "Could we use your oven?" By the end of the day, I had sewed up six ovens. A famous ichthyologist was coming. The coral scientists and their graduate student were coming. The guy who ran the research motorboat was coming. Some masters' students from the University of Sydney were coming.

That evening, we pulped and strained the squashes.

In the morning, we made stuffing and stuffed our fifteen birds.

While I made pies, two students ferried chickens to ovens. In the early afternoon, more students took pies to ovens, and I put four pies into our own.

We had a full house in the cafeteria for our Thanksgiving meal. We feasted on non-cranberry sauce, roasted and stuffed non-turkeys, mashed potatoes made from boxes of flakes, gravy, green beans, and non-pumpkin pie with whipped cream.

It did feel a little odd to celebrate Thanksgiving in the hot tropical springtime. But nevertheless, it did seem like Thanksgiving. After, one of the coral graduate students brought out her guitar, and we sang together, Aussies and Yanks. One of our male students remarked, "The only thing missing is football. My family always watches a football game on Thanksgiving afternoon."

"If you go out to the first coral bommie to the east of the dock," one of the Sydney graduate students said, "you'll find a huge brown sea cucumber with a fat midsection and pointed ends. It has pale stripes on it, too. That cucumber looks a lot like one of your Yank footballs. The water there is only about two feet deep right now. Go have a look."

So, we did. The creature did look just like a football.

The Strangest Fleet

On Heron Island, I was teaching a short course titled Fishes of the Great Barrier Reef, but another of my duties was to supervise the independent studies of several students. While we stayed on Heron, each student was required to do a for-credit independent study, something non-invasive where observations and/or measurements were taken, with a written paper to be graded. Certified divers dove, and snorkelers like me snorkeled.

One student worked with a famous Japanese ornithologist who happened to be on the island at the time, banding all the newly hatched silvereyes (a small bird) and observing their behavior. Another chronicled the nest-building behavior of the noddy tern. I worked with one student, Rob, on egg-laying and parental egg-care of the Great Barrier Reef anemonefish (in more common terms, clownfish), *Amphiprion akindynos*. Several students worked on various aspects of which types of corals were fed upon by which types of reef fish. And so on.

Jo wanted to work on sea cucumbers. I couldn't imagine why. With all the beautiful, active, and fascinating creatures on the island and in the sea nearby, she had chosen sea cucumbers. But Jo was adamant. She was dead set on working with sea cucumbers.

She wanted to record their behavior. This sounded like a reasonable idea. There were countless sea cucumbers near the shore in the lagoon, gliding ever so slowly across the sandy bottom, and for a study, a fair number of creatures are needed to collect data that has some meaning.

The common sea cucumber species at Heron was an almost black, nearly featureless, squishy creature the approximate size and shape of the average banana. This species doesn't bite or sting. It isn't poisonous. Individuals move so slowly that it's possible to keep up with what they are doing. And for some reason, these creatures appealed to Jo, who wanted to find out if they had territories, and if so, how large the territories were. How far did the cucumbers move during a day's time? Did they re-patrol the same areas, or were their movements random? This seemed to be something no one had yet studied.

After we had settled into the research station's buildings and had been given some orientation, it was time to get to work. Eagerly, Jo changed into her swimsuit and grabbed a dive slate from the pile on a table in the kitchen.

The tide was going out when she got to the beach. I sat on the beach rock and watched several of the students, including Jo, beginning several studies in the shallows of the lagoon.

Jo waded out to a place where several of the black cukes lay on the pale sand. She sat in the water (which came to just above her waist) and began to watch. I watched her; she watched them. The tide crept out, and toward the end of the afternoon, gradually made its way back in, until Jo was afloat with mask and snorkel, lying at the surface and looking down at her plump black subjects.

Eventually, when the tide was almost at the full, it was time to head for the big kitchen to cook and eat dinner.

Jo was excited about her research. "Number One stays over by that little *Porites* colony," she told us later over macaroni and cheese and a big salad. "And Number Three didn't move at all. But Number Seven traveled four feet and ended up at that place where there's a piece of concrete near the channel." This sounded promising.

However, her next day's observation was a tale of woe. "Dana, I went out to my study area first thing this morning—and—everybody was in a different place, every single one of them! How can I tell who is Number One and who is Number Seven? They all look the same! They are all the same size except for Number Four, who is little, and I couldn't find him at all." She hung her head over her plate of French toast and sausage. "What am I going to do now? Can I tag the cucumbers or write on them?"

I explained to Jo that tagging them would mean piercing them, and 1. That wouldn't be allowed in a marine sanctuary and 2. If the cucumbers were pierced, they would likely vomit up their entire insides, including their movement muscles, so they wouldn't be having any behavior for a long time after that. And since a sea cucumber's skin is essentially a breathing organ, even if we could find a way to write on them, we couldn't do that without injuring them significantly.

Jo was glum. Her enthusiasm evaporated, and with slumped shoulders, she trudged off through the sand to the cabins. I spent the rest of the evening trying to think of some other study that could be done with the cucumbers, but since they did so little anyway, I couldn't come up with anything.

The next morning, I had laundry to do in the sea water without a super-clean result, but it was better than nothing, and I had some lectures on reef fish to prepare. It was afternoon by the time I got down to the water.

The day was fine and hot, and when I got to the beach, it was almost exactly at the point of low tide. The water in the lagoon was slack and glassy, the air was still, and far on the other side of the lagoon, toward the open ocean, the line of coral that marked the outer edge of the reef was visible just above the surface of the sea.

When Jo saw me, she jumped up from her sitting place in the water and ran to me over the white sand.

"Dana, Dana, I have it figured out!" she shouted, waving her dive slate, all smiles. When she reached me, she said, breathless, "Look out there. See those little floats?" She pointed, so excited that her arm was vibrating.

I shaded my eyes with a hand and, sure enough, I could see little transparent floats here and there in the lagoon, floats with big black numbers drawn on them with marker, orange-topped little floats, perhaps twenty of them, unmoving on the still surface of the water. "What are those, Jo?" I asked her, squinting at them in the bright light.

"They're orange juice containers, little plastic jugs," Jo said in triumph. "Half-pint size. Each one has a handle and an orange plastic lid."

"Where on earth did you get those?" I asked her, mystified. The *Robert Poulson* came only once a week, and she wasn't due for days.

"Well—" Jo hesitated and then told me in a rush. "I went over to the bar at the resort, on the other side of the island. And they have a bunch of these little jugs of orange juice in a fridge at the bar. They use the orange juice in drinks, and they sell the jugs, too. So, I bought some and we all drank them down. I wrote a number on each jug, and then I begged some fishing line from one of the research station people."

"You attached the floats to the cucumbers with fishing line?" I asked. "You didn't hurt the cucumbers, did you?" I asked, picturing the fat black sausages pinched at the "waist" with fishing line.

Jo laughed. "Oh, no. The line is tied to the little handle of the orange juice jug and then I made a big loop around each cucumber, a loose loop. I am watching them closely, and so far, they are crawling along without coming out of the loops. It's great." Her smile was contagious.

"Very clever," I said, but somewhere back in my mind something about the tides was knocking on a door. I felt a prickle of uneasiness.

"How long have the cukes had their floats on?" I asked.

"Two hours," she said promptly. "The fishing line is long enough for the floats to reach the surface. The cukes don't seem to be in any distress."

I wondered how it would be possible to tell if a sea cucumber was experiencing mild distress. Jo went on to tell me about her twenty cucumbers, where they were and where they had crawled, and that she had re-located the little one, Number Four, from the day before.

I could feel the muscles in the back of my neck tingling for no reason. The depths of my brain were trying to tell me something, but I couldn't quite understand what it was.

As Jo continued telling me about her cucumbers, I could see that the inner lagoon current had begun to sweep along the shore. Twice a day, this current faded as the tide went out and then started up once more after low tide, as the tide began to come in.

The tide is coming in, I thought stupidly. *The current is moving right along. The water will be deeper very soon.* But the consequences hadn't quite reached me.

Suddenly one of the other students, further out in the lagoon, began shouting and waving his arms. "Jo! Jo! Your cucumbers!" The guy stopped waving and pointed.

Bobbing swiftly along in the new current, the little floats were heading away from us, out into the center of the lagoon. The water was deeper than it had been fifteen minutes before, and it was getting deeper by the minute.

Deeper! That's what my brain had been trying to tell me. Deeper water! The cucumbers on their short lines would be lifted right up off the bottom.

"Oh, no!" Running, Jo splashed into the lagoon. "Help! Help!" she shouted, and naturally, all our group within earshot appeared on the beach and launched themselves into the lagoon, me included. I slipped on my mask and snorkel and ran out with them, then began to swim.

The orange juice jug floats were moving rapidly now. I finally caught up with one. With increasing speed, the little orange juice float slid southeast on the surface of the water. About eighteen inches below the float, suspended three feet above the bottom, drifted a long cucumber riding in a loop of fishing line, looking for all the world like an overripe banana doing some hang-gliding.

I pushed him from the loop and watched him drift to the bottom, landing in a little puff of sand.

I came to the surface and took a look. The little floats, now madly bobbing in waves whipped up by a sudden fresh wind and still in the grip of the strong current, were heading for very deep water—Shark Bay. Several of us pursued the little floats, swimming like demented seals.

We got to most of the floats and removed the cucumbers. Some of the cucumbers slipped from the loops of their own accord.

But a few escaped us and disappeared into the bay. I often wonder what the cucumbers might have thought with their simple nerve nets and ganglia as they were swept out into Shark Bay. If only they had been sentient, they would have had quite a tale to tell their grandchildren.

Slip-Slidin' Away

It was a cold February in the desert hills of southwestern Idaho, 1982.

I had become a vegetation/small mammal field biology technician working on a grant for the University of Idaho to map plant communities and determine where populations of prey animals lived, so that the boundaries of the proposed Snake River Birds of Prey National Conservation Area could be drawn before the bill creating the preserve could be introduced the U.S. House of Representatives. With one of the highest concentrations of breeding raptors in the world, this desert is sliced into two by the deep canyon of the Snake, and many of the eagles, hawks, falcons, and owls nest in the cliffs there.

On the northeast side of the Snake lay about two-thirds of the proposed sanctuary. On the southwestern side spread the even more sparsely populated, arid Owyhee County portion.

Owyhee County, at 7,700 square miles, is large enough to contain a couple of small states. Its population, fewer than 10,000 people at the time of this story, was (and is) concentrated in several small towns and farmland in the Snake River Valley. The county seat, Murphy, had a population of 97 in the 2010 census. The foothills have few human dwellers and is a good place to be alone.

During the previous summer's field season, my work partner had been raven-haired Vicki— bright, beautiful, and dedicated to the job. This season, we would be split up to partner less-experienced field techs. I would miss working with Vicki. No slacker, she did her share of the work and more, had a quirky sense of humor, plus (something I liked very much) she was eager to learn anything and everything about the desert.

February in this desert is cold and slick with mud. Most of the year's precipitation falls between mid-December and late March. In February, many birds of prey are sitting on eggs, ground squirrel females are pregnant, and it's not a good time to do vegetation sampling because the annual plants are still just seeds in or on the ground.

On this freezing, blustery day, Vicki and I were going vegetation-stand mapping in the Owyhee foothills. Rhonda, a new field tech on her very first day at work, would go with us to learn the ropes.

This would be one of the last days that Vicki and I worked in the field together. We would remember it always.

Squashed between us on the bench seat of the government Chevy LUV pickup, Rhonda was nearly bouncing with enthusiasm. Vicki and I exchanged smiles over Rhonda's head as we drove west down the highway. The sky was overcast, and sleet pecked at the windows. This would not be an easy day.

We crossed the Snake River on the bridge at Walter's Ferry and headed south toward the tiny hamlet of Murphy.

This area was a blank on our vegetation maps. We carried with us a roll of U.S. Geological Survey 7.5-minute maps, on which we would draw the boundaries of the vegetation stands we identified—shadscale, sagebrush, winterfat, greasewood, cheatgrass stands in old burns, perhaps even a tiny remnant of native grasses here and there. Later in the year we would return with clipboards, data forms, decimeter rods, and hundred-meter tapes to sample the vegetation so we could have an accurate record of the composition of each stand we had mapped earlier. In the days before global positioning systems and satellite images, we had aerial photos to help us.

On our map, we found an area near a farm along the Snake that had not been veg-mapped, so we headed down a farm road and then turned south onto a track just outside the fence line. The track was muddy.

Though our LUV had four-wheel drive, it was no match for this valley-bottom glue. Vicki was driving. We churned mud and slid and dug, finally coming to an involuntary stop. We were hub-deep and stuck. Rhonda and I got out and pushed. This had no effect except to splatter us with cold mud. The truck's wheels churned and churned, sending fountains of mud everywhere.

Vicki hopped down from the driver's seat. "We'll have to walk to the nearest farmhouse," she said, glum. She glanced at me, I nodded. We had done this before.

Just as she began to lock the truck, we heard a shout.

"Hey!" came a voice from the nearby field. Sitting on the most beautiful John Deere tractor I had ever seen was a teenage boy, bundled to the eyeballs in Carhartt, gloves, and feed cap. "I'm gonna pull you out," he announced. "Give me a minute to drive around through the gate."

Pull us out he did—and with many thanks, we headed back the way we had come.

"Too muddy down in the valley," Vicki and I decided out loud. "Let's do that unmapped area in the foothills, the hills off that little road behind the Blue Canoe Restaurant." The Blue Canoe, now long defunct, was possibly the most oddly placed restaurant I have ever seen, perched at the bottom of the lowest of the Owyhee foothills, with its parking lot next to a sorghum field, not a house or another business within sight.

We found the dirt road behind the Blue Canoe and started uphill. My turn to drive. It wasn't as muddy here. This was going to be fine.

"Those low, spiny bushes are shadscale," I heard Vicki explaining to Rhonda. "Their leaves are almost round, like little scales. There's quite a lot of it here. Shadscale is one of the plants you need to know."

But by now the road was getting steeper and muddier. What there was of it, was narrowing, growing fainter, and I was growing apprehensive.

Now we were on a steep sidehill and ahead I could see that we were coming to a sharp ravine. If we drove to the bottom, could we get out? The road felt like mush.

"I think we should turn around as soon as I can find a place," I said. A barbed-wire fence paralleled the downhill side of the road. *Could* I find a place to turn around?

Suddenly, there was the edge of the gulch—and the end of the road. I had to stop. If the LUV could have grown wings and flown eighty yards across the steeply carved arroyo, we might have found a road on the other side. Or not. Would it be possible to back up? I never found out.

Gravity and mud took over. Its momentum stilled, the LUV began to slide sideways downhill. The truck slid side-on into the barbed-wire fence and continued to slide. It is amazing, really, how far strands of barbed-wire can stretch before they break.

Steel fenceposts jerked from the ground as we continued to slide, and at last *POP! POP! POP!* all three strands of barbed wire snapped. The truck slid a little further and then came to a stop. At this point, the weather gave up on sleet and began to snow.

We got out.

There was no road in sight, except a faint dirt track about 200 feet down the hill. I got back into the truck, turned on the engine, and tried to turn the LUV's nose downhill, but it was a no go. The wheels spun and spun. Vicki got in and tried. Still no go. By this time, all three of us were abominable mud-women, wet and freezing. My glasses were clotted with both mud and snow.

"Let's push the truck downhill sideways," one of us said, so we did.

I was terrified that the truck would roll, but it slid, with us pushing mightily we finally got it onto the small dirt road at the bottom of the hill.

Vicki drove us past the Blue Canoe, back to the highway. We were soaked and still shivering. "Your face is dirt-brown. And you have a mud ponytail," she said to me.

"You, too," I retorted. "And Rhonda."

Then Vicki did the absolutely unprecedented. "Pie," she said without explanation. She drove south to the turnoff for the town of Grandview and pulled in at the Y Café. Shaking off as much mud as we could manage, we tramped inside—blonde mud-ponytail, black mud-ponytail, and red mud-ponytail—to the astonishment of several truckers.

"Pie," Mud-Vicki said to the waitress. "And coffee."

We savored our coffee and filled ourselves with berry pie as we sat in a booth and watched the snow coming down harder than ever. As I cradled my coffee mug, the feeling began to creep back into my fingers.

"Wow, this is great!" Rhonda said. "Do you go for pie every day?"

Vicki and I burst out laughing.

Toots

My then-husband Eric and two of his students, Brad and Bill, went to a scientific meeting in Florida one year, and afterward, I picked them up at the Boise airport.

"Wait until you see what we brought you," Brad said. My history with Brad made me a bit apprehensive. On an Australia expedition, he had once handed me an epaulette shark.

When Eric's duffel bag came sliding onto the baggage carousel, I had misgivings.

Eric slid the duffel from the carousel and unzipped it, pulling out a muslin bag wrapped in a damp towel. With a flip of the wrist, he unrolled the towel, opened the bag, and slid out something dark and solid. It looked like a small hand grenade.

A wedge-shaped head shot out of the grenade, and I heard jaws click shut. Eric handed the thing to me. "Yipes," I said, narrowly avoiding the snapping jaws. "What??"

"It's an eastern mud turtle," Eric said. "We got a collecting permit and everything, and we thought you would just love to have her."

"Thank you," I said. In spite of myself, I felt a little sunshine glow inside. I love turtles.

I took the turtle home and set up a ten-gallon aquarium for her, with sand, a light, and a haul-out rock. Pleased with myself, I lifted her gently and placed her in the water. She turned and snapped at my thumb. This was the moment that defined our relationship.

I gave the turtle small feeder goldfish, half a dozen at a time. She was ferocious. A batch of fish wouldn't last fifteen minutes with this creature.

I put in a tumble of rocks in an attempt to spread out the feeding a little by giving the fish places to hide. The turtle would shovel aside rocks as large as she was to get to the fish. Even after the addition of the rocks, a batch of fish still wouldn't last fifteen minutes.

After almost no consideration, I named her Toots, because she was tough, aggressive, and thick-skinned. In a bar fight, she'd be the one in there throwing punches, biting, kicking, and pulling hair, not the tender violet needing to be rescued.

When visitors came, I'd amuse them by wetting a finger and touching the glass of the aquarium. Toots would stalk the finger and bite at it behind the glass.

Toots grew and thrived, and put quite a dent in the feeder-goldfish population of the nearby farm store's pet department.

A few years later, she had grown to a respectable six-inch shell length, and her flat abdominal plates and short front claws proved that she was, as we had guessed, a broad, not a dude.

One day I fed Toots some goldfish and went outside to work in the garden. When I came back in, I was shocked; a bright orange goldfish had survived Toots for two hours!

The next morning, Fish was still there. Fish kept well away from Toots, but Toots ignored her. I couldn't believe it. Fish survived and grew. She, too, attained a length of almost six inches, and grew beautifully strong on the tattered remains of her relatives.

One day I realized that Fish and Toots had shared the aquarium for six years.

Every day when I would come home from work, I would hurry to Toots' tank, wondering if Fish had survived the day.

Over the years, Fish seemed to become attached to Toots, and while Toots was resting, would place herself within a finger's width of her friend. Sometimes Fish would draw her filmy tail right across Toots' head. I couldn't believe that my little armored dinosaur wasn't biting that tail right off.

But one day . . .

One day I came home and found Toots' tank cloudy as a storm. The rocks were overturned; the sand had been heaped into one corner.

Through the murky water, I could just make out Toots, a dark lump on the bottom. But after a search, I knew what my heart had told me at first sight. There was no Fish.

Not a single scale had survived what must have been an epic battle. Toots had been fed that morning, so it wasn't hunger that had set her to devour her long-time companion. I can only guess that her basic mud-turtleness had pushed her to the deed at last.

After the death of Fish, Toots went into a long decline, though she never failed to bite at my finger through the glass. I could not find anything obviously wrong with her. I tried turtle vitamins, an internet consult with a turtle vet, and many kinds of food. But Toots continued to slip very gradually down the dark slope that must claim us all in the end.

Two years later, I buried Toots under our apple tree. Perhaps I buried a few remaining molecules of Fish, too. I salute you, tough gals!

Cataviña by Ratlight

Australia was not the only destination for the later College of Idaho field expeditions. In 1984, we took a group of biology students into Mexico on a for-credit, six-weeks-long field trip in midwinter, driving from Idaho through Nevada and California, down to the tip of Baja, across the Sea of Cortez by boat, and back up the coast through Sonora and into Arizona, then back home.

Have you ever shaved the head of a woodrat with a blue plastic razor? On the 1985 field expedition, we did just that.

For weeks, we camped in the Cataviña area, a desert wilderness boulder field of pale granite rocks, some as tall as three-story buildings. The plants there are spectacular—many kinds of big cacti, strange wildflowers, desert shrubs, and the fanciful-looking boojum tree. One of the biologists, Bill Clark, had made friends with a rancher in the area, so we were able to get our water at that ranch a few miles from our camp. Everything else we brought with us. We lived in tents and cooked over a fire on a little gas stove, as was usual on our field trips.

The students attended lectures, learned to identify the plants and animals, and each student did an independent study project. One of the students, Christy, had decided to study the local pack rat, the white-throated woodrat, *Neotoma albigula*, a chunky desert rodent found in the southwestern US and in northern Mexico. Christy wanted to determine how large each woodrat's territory would be, something that had not yet been investigated by anyone.

It was easy to find the woodrat nests. Often occupying the same nests for countless generations, woodrats pile up scraps of vegetation under a fallen cactus giant, in a crack among rocks, or in some other protected niche. Each generation of rats adds a layer of plant material to the nest, so nests can be large and messy. Christy located several nests.

But the rats came out at twilight and went "home" shortly after dawn. How could she determine their territory boundaries in the dark? We'd thought about that. We had brought some dime-like flat batteries and tiny seedlike light bulbs for just such a purpose.

Christy set live traps overnight and caught several woodrats.

The next morning, we shaved the tops of their heads, removing all the fur. At twilight, we attached the tiny light bulbs to the batteries, and glued the contraption to the rats' bald heads. We released each rat at its nest. Christy had found a leaning granite boulder two stories tall, a perfect observation post. She spent the night there, watching the tiny lights darting among the boulders as the rats went about their business. I watched from a different rock. Oh, the solitude of the desert wilderness and the countless stars! The howls of the coyotes! The soft breeze and flickering of the tiny lights in and around the boulders!

The lights faded toward sunrise. The batteries were good only for one night, alas. (We couldn't get LEDs then.)

That evening more rats were caught, shaved, glued, and observed. Christy got a very good study out of this, and the rats were not harmed. (We figured that when their head fuzz grew back, the rats could scratch off the batteries.)

The day before we broke camp, one of the students, Kevin, was exploring the area and found a cave between giant boulders that had tumbled together. He saw a bobcat slip into this cave and emerge on top of a nearby boulder stack. He grabbed a flashlight and followed.

The cave was small, perhaps eight feet high, twelve feet wide, and twenty feet long. Against one of the cave's stone walls, Kevin found a woodrat nest.

This nest, full of dried leaves, cactus spines, and twigs, leaned against the wall of stone and formed a high column of frass, taller than I could reach. I looked closely at the bottom of the nest, on the cave floor. Wait! Were those dusty, colorless little fragments juniper twigs? *Juniper?* In the central desert of Baja? Today, juniper can be found only in the mountains many miles away, and many hundreds of feet higher in elevation. The central desert is much too dry and hot for juniper.

With great care, we collected samples from each layer of the nest.

Once home, we sent the samples to a professor in Kansas. He had studied woodrat nests for many years. We didn't hear from him for months, but when the news came, I was shocked.

Yes, those were juniper bits we had seen, along with pollen of plants not found in that desert since thousands of years before European man had come to the Americas. The Catavina woodrat midden was ancient. Carbon dating revealed that the bottom of the nest was more than 14,000 years old. The professor published his findings in a distinguished journal. But he did not see the tumbled building-sized boulders, the black sky blooming with stars, the flickering lights as the rats went questing in the dark, the great fingers of the giant cacti, the flashlight beams in the dusk of the bobcat cave: he didn't have this memory.

The Heavy-Trap Syndrome

'I've spent considerable time trapping—live trap and release—for various studies and teaching activities.

It's common for students to read, "There are three species of pocket mice in this desert: the little pocket mouse, the Columbia Plateau pocket mouse, and the Great Basin pocket mouse." But holding the small creatures in your hands is the world's best teacher. A drawing or photograph in a textbook or field guide—useful. But for my students, that's not enough.

For many years, in many ecosystems, and for many studies and college expeditions, we trapped small mammals with closed-box metal traps baited with peanut butter, apple, cabbage, and/or oatmeal. The smallest traps were all smooth sheet metal, with a three-inch square opening. These traps were about ten inches long, with no mesh at all, so you couldn't tell what you had trapped until you pressed down the door and looked inside.

Usually, the trap would hold a skittering furry fellow, lightweight and active. But once in a while the trap would be too heavy for such a tiny creature, and I'd open those with care.

In Idaho, a heavy trap usually meant that a grasshopper mouse (gray tiger of the rodent world) had followed the original prey into the trap in order to eat him. Or the small trap would hold a fat chipmunk or woodrat, barely able to squeeze in.

But once in a great while, there was a surprise.

On the 1984 winter expedition into Mexico, we live trapped along our entre trip (with permits). It's wonderful to teach students about cactus mice, and the next morning open a trap and slide one into the net for all to see and handle before letting it go and watching it scurry off into the underbrush.

Learning to identify species is an important tool for a working biologist.Preserved specimens and photographs have their place, but working with live animals is what most of us do and identifying them accurately on the fly is critical. Is this species endangered? Is it rare? Is its population stable? What does it do? When does it reproduce? What does it eat? What needs to be done to keep it going? None of those questions can be answered *if you don't know which species you have.* And for students, seeing those bright beady eyes, tiny pink paws, and bunchy cheeks in a warm, breathing being? Real experience is priceless.

One evening we set our traps in a creosote bush stand on a sloping alluvial fan in the Vizcaíno Desert (central desert of Baja *California Norte*). As the sun came up the next morning, before breakfast, we went out to check the traps. (One should always check traps before the sun heats the metal and causes distress to the little creatures inside.)

I had warned the students about too-heavy traps. Some heavy captures are not calm and safe and must be released with care. One student called out, "Hey, Dana, this trap weighs a ton!"

"Wait," I called, running to him. "Don't open it!"

He handed me the heavy trap. I shook it and heard a dry sliding sound, definitely not the sound of little paws or claws. I shook the trap violently, turned it upside down, pressed the door open, and slung out the contents.

A rattlesnake flopped onto the ground and slithered into a nearby hole. Sometimes snakes will scent trapped rodents, press down the trap doors, and go inside for a snack. Snakes are much heavier than mice!

The students proceeded more cautiously after that. As we caught different species, we began carrying certain traps-containing-contents with us, planning to bring one of each species of creature back to camp to show to those whose turn it was to take down tents and cook breakfast. That day in the Vizcaíno Desert, we caught the Northern Baja deer mouse, the little desert pocket mouse, one of the kangaroo rat species, the white-tailed antelope squirrel, and several others. We finally headed back to camp with only a few more traps to check.

I lifted the last trap. It was hugely heavy, heavy as if the whole thing were stuffed with clay. I shook the trap. Nothing slid. Unlike the other traps we had picked up, the metal was very warm to the touch. I carried the trap back with me.

Back at camp, I had the students stand some distance away, so if the trap contained another rattler, they wouldn't be nearby when it left the trap. I pushed the door open with my finger, pointed the opening away from me, and slung the trap in an arc while hanging onto its closed end. Nothing fell out, and the door snapped shut.

I set the trap on the ground and ran for the toolbox in the truck. Grabbing needle-nosed pliers, I pulled out two of the four long wires that held the sides of the trap together. Then I lifted off the top panel of the trap, and the sides fell away.

Sitting on the floor of the trap was a cottontail rabbit—but this was not just any rabbit. This was a trap-shaped rabbit with square corners and flat sides. Only its fluffy tail looked rabbit-like. The rabbit sat perfectly motionless, the first and only cuboidal bunny I have ever seen.

The students began to laugh.

Jerkily, the rabbit extended one forefoot, then the other. One hind foot shot out to the side, vibrating. Then the other. The ears popped up. After a long moment, the rabbit jumped straight into the air and took off like a shot.

If that bunny had been squashed into that small trap all night, it must have been numb all over.

* * *

The last heavy trap I lifted was in Idaho, in 2002, on the National Guard training area where I was working, in a nearly pristine sagebrush stand we called Orchard Corner.

First thing one bright June morning, my senior technician Jay and I were checking our trap line before the heat of the day.

One of the traps was too heavy. *Hmm*, we thought. *Don't think there are any woodrats in this area. Too heavy for a chipmunk.*

Blurry weasel photo. Weasel launched from
the trap into the air.

We shook the trap. Sound of claws, so not a snake.

Jay pressed the door open a crack and saw tiny eyes and a pointed muzzle. "It's a weasel!" he said.

If you don't know weasels, you might say, "Cute—just press the door open and let it out."

Weasels *are* cute, very. But if you know weasels, you know that this is exactly what you must *not* do. Unlike almost any other mammal of that size, weasels do not fear humans. When you release a weasel from a trap, it might run off. *Or* it might decide that it is a bit miffed and will run up your leg and bite all over your face and neck. That happened to my dad's friend George Castle, who thought a weasel should move off a forest trail to let him pass.

So, we did the sling thing. Jay held the trap over his head. I got my camera ready. I gave him "One, two, THREE," and Jay slung it, opening the door on the downswing. The long-tailed weasel came flying out and I snapped just one photo. An instant later the weasel vanished into the sage.

Be wary of heavy traps!

The Flame Squirrel of Compostela

The mountains near Compostela, Nayarit looked down through tumbled gorges to a neatly ordered, old-fashioned Mexican town. Built in the classical Spanish manner around its cathedral square, Compostela was a beautiful, unexpected surprise.

In 1984, Compostela went about its business much as it had a hundred years before. People went to work, kept house, traded at little shops, lit a candle or two inside the church, and gathered in cantinas at the end of the workday for a cold drink or two.

Around the church square at twilight, young men walked in one direction and young women, usually accompanied by their parents or older women, walked in the other direction. Nods and smiles were given, hats were lifted, small talk was exchanged, and courtship proceeded by nods and winks, the time-tested method that has worked so well, and so ill, for centuries. In 1985, Compostela was a town settled kindly into its ways.

Above the town, short-tree forest and scrub covered the mountains. These were not mere hills, but the Sierra Madre Occidentale, a range of mountains tall and gaunt, clean and parched, threaded with dry streambeds and sparsely populated by people, cattle, and small things of the wild.

We camped several miles from Compostela on our way home from the 1984 winter field biology expedition. It was February, but the days were already uncomfortably warm even halfway up the sides of the peaks.

On that trip, we would stay days or weeks in a good place, and then move on. After we turned north toward home, we spent more of our time on the road. Typically, we would stop somewhere in the late afternoon, spend the night, and move on in the morning.

Compostela was such a one-night camp.

At this point in the expedition, we were travel-worn and weary. We had survived weeks of unseasonably cold weather in central Baja, very hot weather in southern Baja, and most distressingly, a broken axle on our equipment trailer. We were lucky to find a skilled mechanic near Puerto Vallarta. Two more nights would see us out of Mexico and into Arizona, where hot showers, familiar food, air conditioning, and telephones would be available to us once more.

In those days, finding a campsite for a party of several vehicles was not easy in Mexico, even in the western mountains where the population was sparse and there were any number of little dirt roads leading from the highways into the back country. In the USA, there are campgrounds at intervals, places with tables, parking spaces, and bathrooms created just for people who need a place to stay overnight in the outdoors. However, I have never liked established campgrounds, considering them to be bland, tame subdivisions of the outdoor experience, and have used them only when there was no alternative.

But in this part of the Sierra Madre Occidentale range in 1985, there were no campgrounds. Also, it seemed that people were living everywhere. Even the most barren desert flat, when closely examined, hid little houses here and there, and people could be found in every fold of the landscape, no matter how seemingly isolated or inhospitable. It was very difficult to find a place to stay off-highway where the camp would not interfere temporarily with the daily life of someone who belonged there.

It was late afternoon when the expedition swung around a great loop of highway in the mountains and looked down upon the town of Compostela, a small gem with adobe buildings glowing whitely in the sun.

The long light flowed down the scrubby arms of the steep ridges to a valley where there were green, yellow, and brown fields—and every now and then a glint from the narrow, shining river.

In serious search mode now, we followed several obscure dirt tracks off the highway, and after a few unproductive trials, found a grassy clearing hidden in stubby oaks where we could spend the night. But the sun was sinking, and we had to hurry. We needed food.

Stores in rural Mexico in the 1980s were not open after dark. Scrambling back into our vehicles, we swooped down the winding road to town as the sun slipped behind the big ridges of the Sierra Madre. We had to buy food for supper, breakfast, and perhaps for lunch. We planned to drive through Tepic, the capital city of Nayarit, the next day, and there could buy food for the rest of our stay in Mexico.

The dirt streets of Compostela were swept clean, and there were not many vehicles in town in the late afternoon. We found places to park and got out to explore.

I took several students and made forays into the shops. There was no supermarket, and it was too late in the day, or perhaps the wrong day of the week, for the open-air *mercado*. It took some time to glean enough supplies from the small stores to provide three meals for a dozen people.

Our striped jute shopping bags filled, but slowly. In one shop I found five carrots, an onion, and a half-kilo of butter. In another I waited while hamburger was ground from miscellaneous scraps of meat and fat, and I bought a large football of pale string cheese as well. In a third shop I found two hands of short bananas, two pineapples, and a watermelon.

Another shop furnished more carrots, plus a big jicama, green and crisp, and three dozen brown eggs. At last, I found a place with bags of rice and even two jars of jelly. After quite a search, I located a little hole-in-wall place and emerged with four jars (small jars!) of peanut butter, Sanitarium brand, just right for us crazy expeditioners. Finally, a visit to the local *panaderia* netted us several loaves of bread.

Tonight, we would have ground beef and rice stew with carrots, onion, and jicama, and fresh pineapple for dessert. In the morning we would have eggs fried in butter, plus toast and bananas. Lunch would be peanut butter and jelly sandwiches, cheese, and watermelon. Our shopping, at last, was done.

Ducking into the street from the panaderia, we shoppers realized that the sun had gone down.

Directly across from us, the town cathedral reared its spire against the shadowed mountains, and we found that our fellow expeditioners had joined the cathedral-square promenade, walking slowly around and around the square, nodding and smiling like everyone else. It soon became apparent that women and families walked in one direction and unattached men walked the other, clockwise and counterclockwise, a wheel within a wheel, a measured way of meeting and meeting again.

In no time at all, however, it grew dark, so we left the square and drove up into the Sierra for the night.

Soon we had circled our vehicles, built a large fire of oak branches, and filled ourselves with hot food and coffee.

Shining oak leaves reflected the firelight, and great beards of shaggy lichen swayed gently from the knotted branches. These trees were old, and they gave the place an elfin, otherworldly feeling. We felt comfortable in our camp—which had not always been the case on this trip. In the valley below, the clustered lights of Compostela winked out, one by one.

In the near distance, coyotes howled from every point of the compass, and we began to nod over the dregs of our coffee. Very tired before we crawled into our sleeping bags, we fell asleep immediately, so swiftly that it seemed uncanny.

My last thought was that this little clearing must be a place of sleep enchantment, like the glade where Rip Van Winkle had slept his long, long sleep. "This is an enchanted place," I declared fuzzily to the row of students in sleeping bags. A few grunts were the only replies. I drifted away myself. I was so tired that I wondered if I would ever wake up.

As it happened, the next morning I was the first to open my eyes. Before the bustle of breakfast and packing began, I wanted to make the acquaintance of this place, as we were so soon to leave it.

Once out of my sleeping bag, I walked up the narrow dirt track into the elfin forest of the dry side of the Sierra Madre Occidentale.

It was a lovely morning, cool and bright, with the shadows of oak leaves and branches making dappled patterns on the dry winter grass. Overhead a caracara sailed, alert for carrion.

Smaller birds, both familiar and unfamiliar, flitted about, and I used my field guide to identify as many as I could.

Far down, through miles of air, Compostela was awake, too. From the valley below came the occasional flash and blink of reflected sunlight from slow traffic.

Close above, a flock of crows sailed overhead. A small lizard basked on a sunny rock, then flicked himself away. I decided to sit where he had been.

This was a good place.

Flying uphill now, the crows screamed and scattered. I looked that way.

A man was on his way down the track, striding along with purpose, almost with menace. He carried a rifle.

I stood up at once and placed myself in the center of the track, watching him walk down to me. It is a bad idea to surprise a man with a rifle.

As he came closer, I could see the brown face and arms, large black moustache, and blued rifle barrel. This was a strong man in mid-life, and he did not look happy. I stepped toward him.

The last thing I wanted was for him to come upon the sleeping students before I had a chance to talk to him. Frantically I searched my mind for the right Spanish words and phrases. I could passably do well buying things in a market. "*Quiero dos sandias, por favor.*" (I want two watermelons, please.) That kind of thing I could handle, but that was all.

The man saw me and raised his head with a jerk. The rifle came up, too.

"*Hola!*" I said, "*Buenos dias.*"

The man swung the barrel of the rifle toward me, lifted it to his shoulder, and barked out a string of rapid-fire Spanish, none of which I understood. But I could guess that he was saying something like, "Who the hell are you, and what are you doing on my land?" Rattled, I was at a loss for words. My very limited Spanish vocabulary deserted me entirely.

At that moment, a squirrel ran out onto a horizontal oak branch a short way down the track. I glanced at the movement and saw the unexpected, something so beautiful that to this day, my eyes burn at the memory.

The squirrel flung back his long tail and sat up. I gasped. The creature was as orange as if he had been fashioned from living flame. Behind him, the rising sun circled the squirrel in a halo of gold. I let out my breath, made an involuntary gesture.

Then I remembered the rifle. But when I turned to the man, he was leaning it carefully against a tree. I pointed, searched for the right word, any word. "*Linda*," I said, finally, the word for beautiful.

An arc of dazzling white teeth showed beneath the black moustache. "*Linda*," he said quietly.

We stood together and watched the squirrel, in jerky curves, take himself and his impossible color along the branch and out of sight into deep shadow.

Words came back and I began to speak haltingly, "*Estudiantes universidad, Estados Unidos*," I said. "*Una noche. Huevos rancheros*?"

With this, the fellow threw back his head and laughed. "*Soy ranchero!*" he declared, thumping himself in the chest. He was a rancher and by gosh, he would come with me and have some eggs for breakfast! Picking up the rifle, he followed me down into camp.

As it happened, without permission we had camped on this man's ranch and understandably, had made him quite angry. However, once he began to talk with the other expeditioners, some of whom spoke fluent Spanish, he was relieved, then interested, and also quite hungry. It was almost noon before we tore ourselves away from the conversation, packed up the camp, and pulled our vehicles onto the highway to Compostela.

Down and down we flew around the sweeping curves of the mountain highway. The driver of our car frowned at the road, concentrating. Students laughed and chattered in the back seat. Stunted oaks flew past our windows, along with adobe haciendas, laden burros, and cattle egrets standing in a field.

In the valley, the lazy river pooled at the feet of broadleaf trees. A group of women and children waited for the next bus, tucking themselves into the shade at the mouth of a large culvert. A saddled horse rubbed his shoulder on the telephone pole where he was tied. Scraps of cardboard blew along the road's edge.

Other roads joined our highway, and in two hours, we found ourselves in the noon stop-and-go traffic of Tepic, by far the largest city in the state of Nayarit with its hurrying crowds, jangling noise, and pervasive smell of exhaust.

I felt it all slip away as I nodded against the window, finding myself back in the Sierra Madre Occidentale, walking in a forest of oak and mossy lichen, in a place much stranger than it had seemed.

I have seen many kinds of squirrels over the years. Before that day, to me squirrels were ordinary little creatures dressed in gray, in mouse-brown or fawn, in brown-black or even in tones of dry-leaf russet. It had never occurred to me that such a mundane animal as a tree squirrel could come clothed in a coat made of fire. But it was true, one of those unexpected dreams met in the sunlight of a place meant to be remembered.

Sleeping against the car window, I was enchanted, back in Rip Van Winkle's glade, where sleep and waking meet in real dreams, and all the squirrels are made of flame.

Rain

Part One

At a certain time in everyone's life, there comes a grand adventure. This is mine.

In 1984, we were back in Australia with another semester-long, for-credit College of Idaho field expedition.

There are some places in the Outback of Australia where the yearly total of rainfall is zero. That is, it may not rain for years on end. The plants and animals are well adapted to this weather routine. People? Not so much.

On the second of our College of Idaho Australia field expeditions, we felt more experienced, and decided to become a bit more adventurous. Cities, tourist attractions, vacation beaches, shopping? Not important on the agenda. Plants and animals in great variety, unique habitats, wilderness? Definitely on the itinerary. To this end, we had planned to visit far north Queensland, and to get there, we had decided on a route that would take us deep into the Outback desert so we could experience that as well.

Our two vehicles were two rented Toyotas—a minivan and a minibus (not air-conditioned), each pulling a small utility trailer. On this leg of the expedition, we were planning to travel from Chinchilla, Queensland, a town on the edge of the Outback, northwest to Quilpie, where the pavement ended. Then we'd be in the true, deep Outback starting from Quilpie and driving to Birdsville (about 400 miles). After that, we'd go north from Birdsville to Boulia, where we'd find paved road again (about 240 miles, a total of about 640 miles on dirt). From Boulia, we'd go on to Mt. Isa (about 200 miles), and then to the east coast, where we would encounter the tropical rainforest of northern Queensland. We could get gasoline at a few places on the way, and knew we would have to be very careful not to let ourselves become stranded without fuel.

For the past several days, we had been staying at Callitris, an outback sheep station (ranch) near Chinchilla, the home of our friends Chris and Mary Cameron—wonderful, intelligent people who could put up with weird, semi-unwashed college students and the two us for a week without losing their sense of humor or smell.

The students loved following Chris and Mary around the ranch and helping with chores—which, since it was spring, included docking the tails of lambs, notching their ears, and castrating the males. Scaring kangaroos and emus out of the grain fields was a task that some of the students performed with gusto. The students also loved chopping wood, building fires in the boiler-stove, heating large quantities of hot water, dumping the water into the galvanized tub, and taking baths in the kitchen, one at a time.

Once a week, Chris would shoot one of the crop-robbing kangaroos or wallabies to provide food for his working sheepdogs. This was the perfect opportunity for the students to dissect the organ systems of marsupials and learn how they worked. After the anatomy lessons, we would skin the roo or wallaby, salt the hide, and boil out the skull to take back for the college's natural history museum. The sheepdogs, eagerly waiting for us to finish, got the remains.

Chris Cameron was an amateur ornithologist—or, I should say, a true ornithologist without an advanced degree—and introduced us to the many species that nested on Callitris—rainbow bee-eaters, helmeted friarbirds, square-tailed kites—and so many more.

Chris tasked himself with preparing us for our journey across the Outback. He spent hours telling us where we could camp to see amazing birds and where we could find water and fuel. One favor he called in via radio was persuading a far-Outback rancher to let us camp on his property for a night and fill up with water there.

Another thing Chris insisted upon before we left his ranch was to take his shotgun and a handful of shells. "You can return it when you get back to Brisbane just before you fly home. I'll meet you there," he said, determined that we shouldn't leave his ranch without a gun. "I hope you don't need it," he said soberly. We figured that Chris was experienced in Outback travel. We were not, and we should listen to him. We took the shotgun.

The next morning, bright and early, we left Callitris and hit the road, the one paved road that led into the central Outback.

In the little sheep-ranching town of Cunnamulla, we bought as much food as we could carry, then camped off a dirt track several miles west of town in a patch of galvanized-bur, the nastiest, thorniest briars I have ever experienced. But seeing a blue-winged parrot scolding us from a eucalyptus tree was worth every scratch.

Our next camp, highly recommended by Chris Cameron, was at Nine Mile Bore. If we hadn't known about this place from Chris, we'd have passed it by, unaware.

A bore is a well. And this well, just a little way off the main road, formed a deep pool in the desert, a pool twenty-five feet in diameter. The water welling up was clear and sweet. A long pipe carried the overflow to a series of metal troughs surrounded by dirt beaten to powder by the cloven feet of sheep. Sheep tracks were everywhere south of the bore, but we saw no sheep. And no sheep had been in the lovely grove of eucalyptus and wilga north of the pool. We saw no one else at Nine Mile.

The next morning at first light, we set up mist nets at the edge of the pool. We had an official permit for mist netting. It's one thing to see a bird through binoculars, but it is another thing entirely to hold a wild bird in your hands, to look right into that bright little eye, to marvel at the arrangement of feathers and the claw-tipped toes, and to feel the warmth and the flutter of the small beating heart, then to watch the bright wings as the bird flies from your hands. My husband and I were experienced in mist netting; he held a federal permit back home and I a sub-permit, which was why we were able to snag such a permit from the State of Queensland. Great care must be taken in extracting birds and bats from the cobweb-fine nets, but the teaching value is enormous.

The very first bird that snagged himself in the net was a prize—a male mulga parrot, resplendent in shocking orange and turquoise-green feathers. We had a close look, photographed him, and let him go—after he bit one of our graduate assistants. Next, a flock of zebra finches had to be unsnarled from the net, and then a bronzewing pigeon and diamond dove. After a couple of hours, we took down the nets and wandered around the grove of trees, discovering a rare spotted bowerbird in his triangular homemade allee of twigs and grasses that he had lined with white and pale green rocks and bits of glass placed there to impress the ladies.

The students had an exam scheduled for that afternoon, so we spread out tarps in the shade, hauled out our library of texts from the bus, and left them to study.

As the sun climbed in the sky, so did the temperature. It was so hot that I almost flopped down into the shade beside the students, but I was too curious to let slide the opportunity to explore. By myself, I wandered off to the south, following the line of water troughs—and saw some of the most amazing birds of my life.

Variegated fairywrens flitted through the bush and above them, a flock of cockatiels lunched on the flowers of a wilga tree. I heard shrieking as I came upon the last in line of the water troughs. I'd found a bonanza of birds!

A huge pink cockatoo was doing the shrieking, sitting in a bush a few feet above the trough. A male emu brought his flock of babies to drink, a dozen turkey-sized stripy guys with short necks, sleek feathers, and big feet, looking nothing like their huge brown father. Half a dozen small pastel pink-and-blue Bourke's parrots were sipping water further down the trough, along with a flock of bronzewing pigeons and about twenty shocking pink-and-white galahs, the common, chunky parrot of the Outback.

I settled into the shade of a bush—for it had grown very hot— positioned my camera, and waited. Bush flies, the curse of the Outback, began to settle on me.

Chris had told me that Outback parrots and some of the other birds might nest as far as thirty miles from the nearest water

source and would fly the distance for a drink twice a day. I had found their source. Pairs of diamond doves came for a drink, as well as a small group of bluebonnet parrots, followed by a flock of yellow-green budgerigars (parakeets). Then pairs of mulga parrots and mallee ringneck parrots showed up, along with smaller birds, that included the spiny-cheeked honeyeater, black-faced cuckooshrike, the brown falcon, and many more.

Covered in hundreds of clinging bush flies, I watched for hours; the parade of exotic birds coming to drink never stopped. But at last, the shadows of the trees began to grow long across the dry red soil, and I knew it was time to leave my birders' paradise and go back to camp to supervise the students as they cooked the evening meal.

Accompanied by a small cloud of flies, I walked back to camp. This had been a great day, especially since *I* didn't have to take a test! Why on earth would Chris think we'd need a shotgun if the Outback was like this? We were still in the occupied portion of the Outback, few and far between though human habitations were.

We had a lovely campfire that evening and woke to a chilly 38 degrees Fahrenheit. We'd set up the mist nets in the dark, and before breakfast examined and released two broad-nosed bats from the nets. Australia is rabies-free, the ideal place for students to experience bats, which occur there in astonishing variety.

After breakfast, we packed up and drove to the next destination that Chris had recommended, Lake Bindegolly.

Lake Bindegolly is now Lake Bindegolly National Park, designated to protect the fresh and brackish wetlands and the lake itself, very attractive to birds. Bindegolly is a clay-bottomed playa, sometimes containing more than fifty square miles of lake and wetland, sometimes dry as a bone. This year, the lake was not filled to its greatest extent, but much of it did have water. I hear that now, Bindegolly National Park actually has facilities—well, one facility. The facility consists of a short, paved lane to a single paved parking lot. Back then, it was a playa only, just visible from the main road. That was all.

After stopping in the small town of Thargomindah for groceries and gas, we drove the thirty-plus miles to Lake Bindegolly and made camp in the red dunes and acacia scrub away from the shore.

We grabbed our binoculars, bird guides, cameras, and headed for the shore at once. If you are not a lover of birds, this will not mean much to you. But to this day I remember the thrill of seeing such beauties as the blue-and-white fairywren, black swans with rafts of fuzzy gray cygnets, pink-eared ducks, southern shovelers, orange chats looking as bright as mini-Halloween pumpkins on their little stick legs, red-capped dotterel running madly about the shoreline, hoary-headed grebes diving in the deeper water— so many birds.

Finally, the light began to fade and weary, we trudged back to camp to cook dinner. When traveling in the tropics with limits on space and weight, there's no room for ice chests and few opportunities to buy ice, so it's important to plan meals accordingly. Also, trying to buy several days' food for twenty people in a small town can be quite a challenge.

Often, we were faced with strange combinations of ingredients. On this particular night, we fashioned a main dish from rice, raisins, and canned ham. The raisin sauce, while a little strange, was quite good. We christened this dish "Thargomindah Fine Ham Stuff." Because of the bush flies, we ate our dinner holding our plates and walking around camp in a large circle so that the flies wouldn't settle on us or our food. I made up a song about our meal and sang it between bites. I still sing "Thargomindah Fine Ham Stuff" to this day.

After chasing some wild boars away from camp, we soon fell asleep.

The next morning, we headed north for Quilpie, an opal-trading and cattle town near Cooper Creek, the last permanent water for hundreds of miles. We camped along the creek and caught a number of fascinating fish, some of which were species not yet named, but that is another story.

In town, the storekeeper of the very small grocery, whom I hunted up for and persuaded to open the store for a few minutes, helped me buy jam, pineapples, sugar, and mince. While wrapping the sugar in a cone of cleverly twisted blue paper, he asked, "You got plenty of water with you?" I nodded. "Good," he said, "because it hasn't rained between here and Birdsville for about seven years."

Seven years! I thought. *Holy moly.*

I love deserts. I spent several years living in the hills west of Tucson, Arizona, and though the Sonoran Desert could be drier than dry, it rained there every year. Every single year, several times a year. I found it difficult to imagine *years* without any rain at all.

I don't know what it's like now, but in the early 80s, when you traveled the Outback, you paid attention to the police sign-in boards at each settlement on the road, THE road, for there were no others, only a very few short dirt driveway-roads that led to ranch houses. You stopped and signed in every time, because these outdoor chalkboards were the Outback's safety net.

Before we left Quilpie, we signed in on the police chalkboard—how many of us there were, what our vehicles were, where we were going, and how long we thought it would take us to get there. If we didn't show up after a reasonable time, someone would come looking for us. When we signed in at the next chalkboard, 130 miles further west at Windorah, someone there would radio back to Quilpie, and we'd be erased.

There was no more pavement.

We headed out of Quilpie in high spirits, red sand billowing behind our vehicles on the dirt road. Emus stalked away from the road. Families of red kangaroos bounced alongside on springy legs. Squawking flocks of parrots flew from tree to tree. Xeric shrubs and tough grasses spotted the dirt, and vivid red dunes bordered by scrubby trees bounded the wide horizon.

Before we knew it, we had passed through Windorah, the last real town in this part of Queensland. One step away from the road, all was wilderness.

The rules of the road are these: if you see someone coming toward you, you must go two wheels off the road as that vehicle passes you. That way, each vehicle will have two wheels on the road, and neither vehicle will get stuck in the sand, since the highway isn't wide enough for two vehicles to pass. This works well. However, there are road trains. A road train is a semi pulling many trailers, with a big kangaroo-guard made of heavy pipe shielding most of the front of the truck, including the windshield ("roo bars"). If you see a road train that isn't moving over for you, then you must drive all the way off the road to let it go by, hoping you have enough momentum to get back onto the road without getting stuck. The driver will give you a cheery wave!

"Roadkill!" one of the students shouted.

Our vehicles came to a screeching halt. A vehicle had claimed the life of a big male red kangaroo and had done so recently. Batting away flies, we skinned the fellow, several of us working on him. I couldn't find my knife, but a shard of beer-bottle glass found at the side of the road made an excellent substitute. We salted the hide, rolled it up, lashed it to the rack on top of the station wagon, and went on. Fifteen minutes on the road west of Windorah, and we had gained a fine specimen for the museum back home.

The land was dry, dry, dry, but there were trees scattered across the landscape. This seemed very odd to me and still does today.

We saw a number of road trains but few other vehicles. Our destination for the night was Cuddapan. The ranch was owned by Chris Cameron's friends Herb and Pearl Rabig, a large station in the middle of, well—nowhere.

We camped out of sight from the road and had time before dark to set up mist nets over a stock-watering tank a little distance away. In the morning, we found we'd captured a mastiff bat, a species not known to occur far out in the desert. Beth, the student whose project this was, was ecstatic. After we returned to Brisbane, we discovered that this occurrence had extended the known range of the species by over 500 miles!

Taking water on board at the ranch, we bade farewell to the Rabigs and drove on.

About midday, we arrived in Betoota, a "town" which was essentially just a hotel—an old hotel. The bush flies were everywhere, clinging to everyone. By now, we had learned that we had two choices: 1. To shake our shirts, shake our heads, and bat at the flies constantly. Or 2. To let the flies settle on our backs and hats and stay there, hundreds at a time, harmless but unnerving, and every minute or so, do the one-handed "bush salute" to keep them from settling on our faces. At least the bush flies didn't bite.

At Betoota, we had hoped to buy whatever food might be available, but nothing was to be had except a few cold Cokes from the hotel's cooler. Yes, we bought them all from Betoota's population of one person. Something cool, *ahhhh!*

Guzzling the last of the Cokes, we left Betoota.

We had just over a hundred miles to go to reach the next town that was a town—well, sort of—Birdsville. We figured that we'd get to Birdsville by late afternoon, buy what food was there to buy, and then would turn north on the road to Boulia, finding a camping spot with a couple of hours of light to spare. That was the plan, but as we drove out of Betoota, we noticed low clouds on the western horizon.

And the adventure began.

***

We were now in the deep desert, with very few trees and shrubs. Between the low hills appeared large areas of gibber, or stones, all beveled flat by the scouring of sand in the wind, smooth and glossy with the mineral varnish that was the result.

A road train came toward us and passed, the driver waving vigorously. *We must be the first people he has seen for some time,* I thought. We continued on. The western clouds sailed toward us, and wild wind blew dirt in blinding drifts.

Then it rained.

With Betoota many, many miles behind us, it began to rain hard. "Wow, we are experiencing rain in the far outback," we said to each other at first. "Cool! Not everyone gets to do that."

it continued to rain—pouring, drenching rain, driven by strong winds, our windshields plastered with dust-laden water, running rivulets of mud down the glass.

Mud. The road turned to mud.

We thought this was fun, doing a little slip-sliding on gradual turns, sledding down the gentle hills, powering our way up again. But the thin skin of mud became deeper mud, became long, deep mud-bottomed puddles on the flats.

The hills began to have a menacing look. Wheels spun as we held our breath; then they would catch, and up we would go. My hands would grip the back of the seat in front of me, and I found myself whispering, "Please, please," as we started up each hill.

Finally, there loomed ahead a short but steep hill. The bus took a run at it, followed at a distance by our van. We spun, we wallowed and slipped—but we made it to the top. My husband, who was driving the bus, stopped us for a breather.

At the top of this hill, having come from the other direction where we were heading to, sat a road train, splattered to the roof with mud. The driver, holding his wide-brimmed hat to his head with one hand, got out of his cab. Our two drivers met him in the middle of the sticky road. I opened my window so I could hear what Mr. Road Train had to say.

"It's bad where you're headed," he said, "and it's going to get worse in a hurry. You ever been in an Outback storm?"

Our drivers shook their heads.

"You hit the signboard at Betoota?" the road train driver went on, shouting over the wind.

Yep. We had hit the signboard at Betoota.

"OK, here's what you do. You find a place, right away, where you can get all the way off the road, *not* where there's a watercourse or low place. Got it?"

Nods.

"It might be good to find a place where there's some puddles not too far off, so you'll have water," he continued. "I have water, so I'm going to stop right here. The police will send a plane over in a day or two to see where everybody is, to see if we're OK. If you don't have food or water, you can signal the plane and they'll drop some. There's a good place in some dunes a few miles back down the road, if you can make it there—sand doesn't turn to mud. Or you can stay here with me. I think we're going to be mudded in for a while."

We thanked him, waved, and went on.

It was touch and go, visibility limited by the torrential rain, the ever-deepening mud slowing us, the engine of the bus laboring, the wheels churning.

Light was fading toward the end of the day, and in the tropics, light fades fast. We made it down the last hill in a controlled skid and wallowed jerkily across some flats, thunder booming overhead. It wasn't fun anymore.

Just when we thought we couldn't go forward another inch, someone shouted, "Look! The dunes!"

And there they were, low dunes immediately to the south of the road, just on the other side of a dry wash, with—amazingly, a flat place twenty feet off the road, a flat place in sand, where we could park.

We pulled both vehicles off the road and stopped. The rain was just a sprinkle now.

I climbed out of the bus and looked around. The road stretched to the vanishing point east and west, filled with water, empty of vehicles.

This world was pale orange sand, orange-brown and gray mud, puddles, two dry washes, scattered shrubs, a few spaces floored with fist-sized rocks, bush flies, fading light—and us. Nothing more.

Hurriedly, we tramped over wet sand and set up our canvas tents on the flat area—girls' tent, boys' tent, professor's small tent, and cook tent. We stuffed our sleeping bags and duffel bags into the sleeping tents and set up our gas stove in the cook tent, fixing a quick meal and leaving the dishes for morning.

As I began to zip up the door to our tent, an ominous flickering to the west caught my eye. More rain would be coming in the darkness.

And in the darkness before dawn, it reached us, another storm. It rained as it can rain only in tropical latitudes—buckets and waves, waves and buckets of water—lightning and thunder,

thunder and lightning.

I huddled motionless in my sleeping bag, careful not to touch the canvas walls of the tent lest I break the air seal within, knowing that if I zipped open the door, the inside of the tent would be drenched in a moment.

Our camp in the dunes, the morning after the rain.

Dawn crept in, and the first light of day that bled through the pale green walls of the tent made me feel as if I were underwater, submerged in a bubble of air.

Eventually the rain fluttered to a stop, and I heard muffled shouts. "Aaaaaaa! Aaaaaaa!"

I struggled into my jeans and pushed my way out of the tent.

Our little tent was still standing. The girls' tent looked fine. But the boys' tent? The boys' tent was a collapsed, wrinkled canvas rug, not a pole still upright, with pools of water between the lumps that were the boys. These lumps were struggling. "Aaaaa!" came another shout, followed by curses and "Where's the door? Get off me!"

One of the guys, though he got himself soaked in the process, had bailed from the tent during the storm and had spent the rest of the night in the bus. The others had opted to stay in the tent, but the wind and rain had proved too much, and the tent poles had toppled.

One by one, the boys, damp and bedraggled, crawled out from under the flattened canvas. Everyone was fine; no one had been hurt by the falling poles as I had feared. The cook tent had also survived the storm.

It was now time to assess the situation—the world had changed overnight.

The low places between the small dunes were now ponds. The two dry washes had become deep, fast-running creeks of muddy water, flowing into a wide, brand-new lake. At the bottom of that lake somewhere, was the road. I took a long look all around. This was definitely a fresh take on "marooned on a desert island."

I might note that this was an outstanding group of students—bright, kind, honest, motivated, hard-working, funny, and curious. Having lived with them for several weeks, I knew what they would do.

Faced with this hardship, that day's set-up camp committee, heaving waterfalls from the fallen tents, set it up again. Sodden sleeping bags and clothing were dragged out and draped over bushes to dry. The breakfast committee dove into the cooking tent and soon served up a hot breakfast, then dipped water from one of the ponds, boiled it, and did the dishes. The students who were on the food-buying committee were, of course, idled, so they explored the dunes and gathered wood for a campfire. Then they hauled out the live-traps and placed them near some burrows they found, to be set in the evening.

By noon, the clouds had left us, hurrying east to wreak more havoc upon the great east-west road.

After camp was secured, we explored our watery world. I found that away from the dunes, the soil was very clay-ey. Flat places that looked like dry soil proved to be deep glue when I tried to walk there, so I stuck to the dunes. We decided that it was time for a little birdwatching.

The first species I noticed was the little corella, a lovely white parrot with a blush of pink around its blue face. Then came the red-backed kingfisher, going after lizards rather than fish. And here we saw the astonishing crimson chat, a red-letter bird in anyone's bird list, not to mention the rufous songlark, hooded robin, and rainbow bee-eater. Galahs were here, too, improbably pink in this mud-orange world. A pair of brolga cranes got up from one of the dune ponds. We found several kinds of eucalyptus trees, the paloverde-like *Frankenia*, strawflowers, and spinifex grass. And we found coolibah trees. Of course, we had to sing "Waltzing Matilda." Everybody knew "Waltzing Matilda." But we saw no jumbucks, also known as sheep of the male variety.

Bush flies came out with a vengeance. Typically for lunch we would set up one of our folding tables, and we'd lay out pieces of fruit, bread, peanut butter, jelly, jam, knives, cheese, and lunchmeat so that people could make their own sandwiches. But the thought of jars of peanut butter and jam gone black with crawling flies was not appealing. So, we set up the lunch table inside the cooking tent. We slid into the tent, made our choices, grabbed a sandwich and a piece of fruit, and slid out. Only a few flies got in.

We spread a tarp on the damp sand in the spidery shade of a *Cassia* and had afternoon lectures. Then the cooking committee got to work on stew and coffee, and the trappers trekked into the dunes to set the livetraps.

As we sat around the campfire in the dark that night, the stars winkled brightly high above—far, far away from any sources of air or light pollution, less numerous than they appear in the Northern Hemisphere, many constellations unfamiliar. I looked up at the Southern Cross and had a sudden thought. I said aloud, "What is the most striking thing about today, the whole day?"

One of the students answered, "We haven't seen a single vehicle all day," he said. "There are only two transcontinental roads in Australia, and this is one of them, a major highway. And we haven't seen a single car or truck on the road."

"Yes," I said. "That's what I was thinking."

In the wee hours of the night, I awoke, unzipped the tent door, and crept out in a t-shirt, shorts, and bare feet. My toes dug into the cool, wet sand. A pale half-moon, sharp as if cut in half by a knife, sailed overhead.

Far to the south, a howl, low and mournful like a wordless requiem, came to me from behind the dunes. Another howler, much closer, answered. *Dogs*, I thought. I knew the howls of coyotes and wolves. These were the howls of *dogs*. But what were dogs doing here in the far Outback? Then I knew. Dingoes.

A little chill shivered down my spine. "In your world," I whispered. "I am just a traveler."

The Australian aboriginal peoples believe that strange creatures lived in a time of magic and great adventure before the world of today. They call that time the Dreamtime. I had thought the legends fanciful and entertaining. But in this remote wilderness, as I listened to the wild howls of the dingoes, I felt anything was possible—a rainbow serpent, a crocodile man, shape-shifters, and above all, spirits of place. After a long time standing and listening, I returned to the tent and sank into sleep.

***

The morning dawned sunny and bright, giving us a number of very pretty long-haired bush rats, *Rattus villosissimus* (meaning "the hairiest") caught in our livetraps. We found dingo tracks in the sand all around our camp; the wild dogs had come to investigate while we were sleeping.

The two creeks had stopped flowing and had transformed into strings of disconnected puddles. The new lake was gone, too. The road, though full of puddles, was visible now. Today was cooler with a brisk wind. The landscape was drying out rapidly.

To the west, a line of mare's-tail clouds danced above the horizon. Another storm front was on its way.

Not long after breakfast, we heard the sound of an engine and soon spotted a small plane buzzing toward us as it followed the road from the air. Two of the students quickly wrote a large "OK" in the sand. We waved as the plane circled overhead, wagged its wings, and then headed further west. The police were indeed checking on everyone listed on those chalkboards. That was a good feeling.

We found ourselves glancing at those mare's-tails over and over. Finally, someone gave voice to what we were all thinking. "Let's go on," someone said. "If we get stuck, we'll be on the road. We'll be found." A tiny, barbed voice whispered in my mind, *Yes, but if we get stuck, will we have to camp in the mud?*

We nodded among ourselves, and without a definite decision, spontaneously began packing up. The tents were dry, the sleeping bags nearly dry. After all, what did we have to lose?

By 9 a.m., we were on the road. Someone shouted, "Birdsville, here we come!" and we were off.

I found myself in the van. Mike Shannon, another of our graduate assistants, settled into the driver's seat.

The road was abominable—mudhole after mudhole. The mudholes were never-ending. For most of my life, I had regularly experienced barely drivable, sometimes dangerous little roads and tracks, from central Mexico to arctic Alaska, and never had I seen a road this terrible.

The secret to driving in mud is *momentum*, a dangerous dragon to control. The van blasted through every mudhole, following our bus and its trailer. The road was under water much of the way—sometimes continuously submerged for hundreds of yards at a time.

The main East-West highway in central Australia in 1984,
with our group pushing the van.

Beneath the thin reddish surface, the desert had turned into a sea of thick, gray clay. Mike was an excellent driver, and steady—a good thing, because there was a lot of screaming going on in that van, and I was doing my fair share.

After a few miles, we saw a vehicle coming! The little 4x4 Toyota kindly scooted off the road so we could stay on it. We waved happily and blasted on.

The trip to Birdsville couldn't have been much more than thirty miles, but it took hours. We passed a few more vehicles, all 4x4s. As we splashed and shimmied past, one driver shouted, "This road is closed if you don't have four-wheel drive!"

We rolled into Birdsville at about two in the afternoon.

Birdsville, the home of a popular racing event every September, had gone back to sleep.

In 1984, the town had a post office, small travel trailer park, police station, hotel, a scattering of homes, and about 100 permanent residents. Every year during race week, five to ten thousand people, many arriving by plane, would flood into Birdsville to watch the racehorses do their thing. But now, in the first week of October, the race tourists were long gone.

We pulled off the road near the trailer park to wash the mud from our vehicles. By now, the windows were almost uniformly covered. Our bus, van, and both trailers had turned gray with mud splatter and were still dripping globs of mud into the dirt.

A man sitting beside his camp trailer glanced at us and did a double take. He rose abruptly, leaned over the trailer park fence, and said with strange emphasis, "It's you!"

"I guess so," I told him, taken aback.

"Yes, but—it's *you*!" he repeated. "I mean, you Yanks—you made it!"

Now, I was curious. "What do you mean?" I asked him.

"You're that Yank college group, right?"

I nodded. How could he possibly know this?

"Lady, you have been all over the national radio news!" he said. "And I'm the first one to see that you made it! Wow!"

Wow, indeed. A small part of me hoped that our students' parents and the college administrators hadn't been tracking Australian national radio news.

The next thing I did in Birdsville was to raid the grocery store. Its "groceries" consisted of two tiny jars of peanut butter and a dozen or so dusty cans of various vegetables. I didn't care what kind of food was in the cans. I bought them all.

The students very much wanted to continue southeast of Birdsville to the "three corners," where Queensland meets South Australia and Northern Territory. It was only a few miles out of our way, and the rain at Birdsville had been very light. The grocery store owner told us that he'd had word that the three corners road was still good. We loaded up.

After setting foot in all three states and taking a formal portrait of everyone in front of the sign, we returned to the serious business of getting to Boulia. We hurried back to Birdsville, signed the police message board, and began our journey north.

Staying in Birdsville until the road dried out was not an option. Road trains with their vital supplies were still stuck on the road, wherever they happened to be when the world turned to mush. There was not enough purchasable food in Birdsville to feed us even one meal, and we only had two days of food left in our supply trailer.

From Birdsville to the next "town," Bedourie, is about 115 miles—but in 1984 Bedourie was a settlement of ten houses or so plus one gas station, not a town in the grocery store sense of the word. The next *real* town would be Boulia, 240 miles north.

However, our spirits were high. Hadn't we defied the mud and made it to Birdsville? Hadn't our journey been all over the news? We were badass, we were. Nothing could stop us!

Besides, the road was drying out rapidly. It was somewhat disconcerting to see a hand-lettered "Road Closed" sign at the turn to the north. But a closed road? Hah. We'd been there, done that, and were ready to do it again. Besides, we reasoned, all we wanted to do on this day was drive for a couple of hours and find a place to camp. We passed the "Road Closed" sign and went on.

At first, the road was not bad at all, compared to what we had gone through that morning.

I kept a lookout for a group of several hundred tall trees that I had read about before leaving home, an isolated remnant of a great post-Ice Age forest now barely hanging on in the depths of the central desert. Waddy, the trees were called, *Acacia peuce*— and we found them growing just east of the road, strange and out of place, looking like a gaunt band of ghostly nomads straggling south in search of a better place to live.

We drove on. And then the road changed.

We soon discovered that the road north was even more badass than we were. The farther north we drove, the worse the road became.

We saw no one. To the west, the mare's tails kept gliding

toward us, with unbroken cloud cover behind them. This did not bode well.

Vegetation thinned out and almost disappeared. The desert was flat now, or nearly flat, floored with the sand-polished rocks—gibber—stretching away empty on all sides to the horizon.

Mike was again at the wheel of the van, and we followed the bus, once more blasting through mudholes, wheel-spinning and skidding, hoping for a better road up ahead.

Soon, those of us in the van began classifying the mudholes. I suppose that was a natural consequence of being biologists. An easy, shallow, short mudhole was a Class One. Class Two's were longer and deeper, and required some skill to navigate. Class Three? You could get stuck in a Class Three, but usually could power your way out. Class Four was serious—long and deep. We got stuck in several Class Four's and had to get out and push. But we got through them.

Another of our graduate assistants, DuWayne, being fond of tunes, had a pocket tape recorder, and he recorded several minutes of "Eeeee!! Watch out, Mike! Class Three, Class Three coming up! Aaaaaaa! Splash!" from inside the minivan.

The mudholes grew longer and deeper. Ramming through them began to seem dreamlike, unreal.

Class Twos became a thing of the past. Everything was now Class Three or Class Four. All of us were splattered with mud from pushing the vehicles out of the mudholes. We got out again and again, pushing our vehicles free of the horrid, clinging mud.

But as the afternoon grew old, as you might have guessed, we met a Class Five.

The bus met it first, wallowing in the deep mud. This mudhole was about fifty yards long. Mike brought the van to a stop some distance back, and we all got out to help push the bus. My journal tells me that the mud there was fourteen inches deep. Yep, I measure things.

We pushed ourselves silly, and the bus would not move. Our legs below our shorts were gray with mud and our arms caked to the elbows. The sun, though westering, beat down on us with hundred-degree menace; and we were getting very tired.

Then someone gave a shout. "Snake! Snake!"

Part Two

I n Australia, a snake is not an animal to take lightly. I ran to the east side of the bus, where someone stood pointing.

Yes, it was indeed a snake, a thick, featureless fellow over six feet long, arrowing straight toward the bus. Someone asked, "What is it?"

I think most of us knew as soon as we saw it. I had drilled the students over and over on how to identify Australia's venomous creatures. This was an eastern brown snake. Sounds innocuous, doesn't it? Just a brown snake. The snake was brown, all right.

My heart sank. The snake kept coming. "Throw rocks!" one of the students, Dave, said, and yes, that was a great idea. If we could get the snake to return to the open desert and leave us alone, that would be best. We pried gibber from the mud and heaved them.

But the snake kept coming closer, perhaps seeking the shade under the bus. And we were stuck. There was nowhere to run except into the stiflingly hot vehicles or out into the muddy desert. And what would happen if the snake crawled under the bus? What if it wound itself into the undercarriage, as I had seen snakes do many times, and struck at us when we tried to go in or out? Eastern brown snakes can be very aggressive, and this one was not discouraged by the hail of small rocks. The snake was now within twenty feet of the bus and still coming.

My husband, sitting in the driver's seat of the bus, saw the snake, loaded the shotgun, and called for Brad. Through the bus's window, he passed the shotgun to Brad. Brad took the gun, turned, and fired. His aim was true.

The brown snake, now mortally wounded, writhed and writhed on the scattered rocks of the desert floor. I remembered to breathe again.

The eastern brown snake of Australia belongs to the cobra family. It's one of the two most venomous land snakes in the world, second only to the taipan. The brown is what the Aussies call a "one-step snake," meaning that once bitten, you don't get far before you die.

Even if we hadn't been stuck in the mud—in the far Outback, if someone had been bitten, we'd have had no chance, no chance at all, to get to help in time.

Brad darted forward, picked up the still-writhing snake, and handed it to me. "Now you can get him pickled," he said, and went off to put the shotgun away.

I struggled with the brain-dead but still reflexively biting, heavy snake until I could get it unwound from my arms and safely

Brad with the eastern brown snake.

dropped onto the ground.

I then went to get the can of formalin from the bus's trailer. Mr. Brown would fit nicely through the opening of the five-gallon metal container and would be a valuable specimen for the college's collection. I made some slits in the snake's belly so the formalin could do its work. Then I wrote a paper label, attached it to the snake's jaw with string—after I was certain that it wasn't going to move again—and hand over hand, fed him into the formalin can's opening, screwing the lid on tight. Bye-bye, Mr. Brown. I took a moment to close my eyes and bless Chris Cameron for insisting that we take his shotgun.

Someone got the bright idea of unhitching the bus's trailer, and without the extra weight, we were at last able to push the bus out of the Class Five. We had to push the van as well, since it had stopped at the edge of the mudhole behind the bus.

At last, at last! We fired up the vehicles and went on, sweaty, muddy, exhausted, and triumphant. We had conquered a Class Five.

We powered through more mudholes but saw no more Class Fives that day.

Though we were in the deep desert, the road began to pass through an area of little drainages and wide stretches of rushes and sedge, filled to the brim by the recent rain. The marshes were alive with birds—pink-eared ducks, black-footed dotterel, zebra finches, and more. Bustards rose on huge brown wings as we passed. Grey falcons flew overhead on their final hunts of the day. The sun began sliding into the far western horizon. It was past time to find a place to camp.

Just when we were about to despair, we found a good place. A line of little dunes undulated to the east of the road, and at the foot of the closest dune, a pond some fifteen feet wide and twenty feet long mirrored a border of sparse, scrubby trees. Best of all, here the road crossed an area of sand, wonderful sand, and the area between the pond and the road was a flat place of paler sand. This was the place!

We set up camp near the—yes—the billabong, not far from the nearest coolibah tree. And we sang that song as we worked, though again, we saw no jumbucks. By now, we had truly experienced the Outback as few have experienced it and felt that we could certainly be called jolly (if not swagmen). One by one, we went down to the billabong and washed ourselves free of mud. We even had enough energy to set out a few live traps on the closest dune and put up a mist net at one end of the billabong. We were in luck—somehow the clouds had fled to the horizon, leaving the sky clear and starry. We heard bats chittering overhead in the dark.

I woke up in the morning twilight to the faint howls of dingoes fading on the wind.

As the day brightened, I was alarmed to see the sky clouded over and threatening rain. We had caught no bats or birds in the mist nets. Only a few long-haired rats had entered the livetraps. Hurriedly we packed up camp. It was time to hie ourselves out of the Outback.

We got back onto the road, which was still muddy. And swampy. Class Three's, Four's, and Five's—abounded.

We soon got the bus stuck once more, in a very bad place. Fiercely determined, we eventually pushed it out. And a few miles further on, got the bus stuck again. And pushed it out.

Part Three

Then, on the dashboard of the bus, a light came on. To our dismay, this was the oil light. I remember very clearly when I was learning to drive at thirteen, my dad saying, "Danny, see this oil light? This is a very serious light. If you see the oil light turn on, first turn off the key. Turn the engine off before you even think. Then worry about what to do. I mean it. Otherwise, you will burn up the engine." The driver turned off the key.

In smashing through mudhole after mudhole, we had loosened the plug on the oil pan, as one of the students discovered by crawling under the bus. Now the plug was gone, lost in the mud. The bus was utterly empty of oil, and we hadn't thought to bring any cans of motor oil with us, something we never forgot again!

Truly, there was only one course of action. We knew from Chris Cameron that there was a gas station at Bedourie, and we figured that we were about twenty miles south of the place. The only thing to do was to send the van on to Bedourie and, hopefully, come back with some oil for the bus.

Mike Shannon drove, and I was already in the van—along with several of the students. Off we splashed.

The road was horrible, and after a few minutes, no one spoke. Twenty miles of mud, and Mike got us through them all.

Bedourie was a mere handful of houses, but in what passed for the center of town, there stood a gas station, complete with an antique gas pump. And the gas station was closed. "Well, why not close the station?" I told the students. "It's likely that no one has passed through Bedourie for the past three days."

We began knocking on doors, starting with the tiny police station. No one was at home there. Eventually we found the woman who ran the gas station. "The police have gone up the road to see how it is, and whether anyone is stuck and needs help," she told us. "But you came from the south?" Wide-eyed, she shook her head.

She opened the station for us—but there was no oil on the shelves. The woman jammed her hands into her pockets, bowed her head, and said, "Now, let me think." Then she strode off down the road, doing some door-knocking of her own.

We sat on the ground beside the van, batting at bush flies. "A person can live for a month without food, but he has to have water," a student remarked. "Well, there's lots of water." No one spoke again for some time.

Overhead, dark clouds gathered.

Then, in the distance, I spotted a man carrying a bucket. Closer and closer he came, and when he noticed us, he began to wave.

The grimy plastic bucket he carried was full of dirty oil, oil that was black as coal. "This is all I've got," the man said. I restrained an impulse to hug him until his eyes popped. "It's used oil," he went on. "I just drained it out of an old rig I'm not using. But you are welcome to it. This should be enough to get your little bus back in business."

Thanking him profusely, we set the bucket on the floor of the van, and with optimism that was a bit forced, told the gas station woman that we'd be back soon to fuel up. Grim-faced, we steeled ourselves for the difficult journey back, and headed south.

May I say that, even if you are down on the floor trying with all your might to hold a bucket still, it's difficult to keep all the oil in the bucket when you go rocketing through hundreds of mudholes? We took turns holding it, and we got back to the others with most of the precious oil still inside that old blue bucket.

While we were gone, one of the students had found the oil pan plug, which had been glued to the bottom of the bus with mud. We used the funnel we had brought for pouring fuel into the Coleman stove and filled the bus with that beautiful, filthy Bedourie oil. The bus started up, and we all cheered. A stunning letter-winged kite, startled by the noise, got up from the closest dune and sailed away.

Both vehicles needed fuel. Through many bogs and mudholes, we struggled back to Bedourie. The gas station lady we found was in a different house this time, but she went to the station and cheerfully filled up the bus and the van.

Above us and to the west, the sky had turned deepest indigo.

We headed north out of Bedourie, hoping to beat the storm. To our delight, the road here was better, with some low places concreted in and others filled with gravel. Still, there were silty flats, drying now, that would have become impassible after five more minutes of rain. We drove as fast as was safe.

The sky to the west and south darkened to near-black and drops of rain spotted the windshields.

And—here came a tilly (a pickup-like small truck)! The driver pulled over to let us go by and shouted out his window. "How's the road!"

Mike shouted back, "Nasty!" The tilly, which, enviably, had four-wheel drive, disappeared south into the looming storm.

In spite of the sprinkles of rain, the ground was drier here, and though we encountered more bogs and mudholes, in many cases we could leave the road and drive around them. We got stuck, briefly, only once.

Once, we detoured around a huge king brown snake stretched across the road—venomous and deadly, but not as deadly as the fellow who rode in our trailer, sloshing in a bath of formalin.

Coming up was a deep watercourse, and we were thrilled when we saw that there was a bridge over it, an actual bridge. We met a few more 4x4 tillies heading south. Slowly, we were leaving the Outback wilderness, coming back into civilization.

The sky continued to threaten us, but aside from occasional sprinkles, the rain held off. The road continued to improve and eventually became a real gravel road, then an actual paved road.

In the late afternoon, we drove into Boulia, a town in every sense of the word.

Boulia had a gas station, and the gas station had water hoses. Here we washed our vehicles and trailers. This took some time, as layer after layer of gray, red, and tan mud had dried onto everything. The mud came off in chunks. One of the students estimated that the bus had been carrying as much as 200 pounds of mud on its undercarriage.

Boulia also had a café. Though eating out was not in the expedition budget, we made an exception, and soon everyone was munching through big hamburgers complete with slices of beet or pineapple and topped with fried eggs.

Boulia had a grocery store, too—a great grocery store, a wonderful grocery store, a brilliant grocery store for such a small town. As I and that day's food-buying committee pushed our carts through the narrow aisles, picking up cereal, carrots, potatoes, ground beef, pineapples, and bananas, the other students were busy replenishing their personal supplies of chocolate. All the chocolate we did have, had melted in the desert heat, and had oozed from its wrappers during the past few days, leaving sticky messes in daypacks, purses, and pockets.

In this little grocery store, the meat, dairy, and chocolate each had separate refrigerated lockers, because, of course, the store itself was not air-conditioned. You opened the half-sized wooden door so you could see in, made your choice, and latched the door.

Refrigeration? I had to feel it. The cool air on my face was intoxicating, and I got a stern look from the store clerk when I kept the meat products door open too long. One of our students, Julia (now a distinguished paleontologist), was seriously foraging for chocolate in a bottom locker—so seriously, that only her legs stuck out as she bathed in the cold air. Envious, but seeing the clerk giving me another stink-eye, I enlisted the help of students Laurie and Beth, and we pulled Julia out by the ankles.

Happily, Boulia also had a caravan park (a private camping park). The park was clean, grassy, and sported real bathrooms and showers. We drove in not long before dark and set up our tents among scattered eucalyptus trees.

Sometime during the night, I woke to hear the patter of rain on canvas, a gentle rain. Boulia was getting only the thin, ragged edge of the evil black storm that had dogged our rearview mirrors and made us anxious all day on the road.

We had signed ourselves free of the far Outback on Boulia's chalkboard.

We had food; we had gasoline; we had oil; we had chocolate; and we had a paved road ahead for tomorrow. And we'd had an adventure. I turned over in my sleeping bag and let the rain sing me to sleep.

This story is dedicated to Mike Shannon. May he rest in the profound peace of the Dreamtime.

The Monster

October 1984. The College of Idaho's semester-long Australia Expedition was camped near Ravenshoe (that's Raven's-hoe) in the montane tropical rainforest, and for our first time in this forest, we were planning to go possum-spotting at night.

That day we had done some scouting and found a path along the edge of an abandoned field bordering the rainforest. We figured that for our first night venture into this forest, a path along the edge would be safer than blindly walking into the dense trees. We should get to know this place a bit, first. We'd also be here for a week.

Before we had left the College of Idaho, I had lectured and tested the students on recognition of dangerous animals and plants of Australia. My fondest wish was to bring them all (the students!) back alive. I taught them about deadly snakes, stonefish, sharks, funnel-web spiders, box jellyfish, and other dangers, hoping I hadn't forgotten anything important. The students did well on that first test, and I began to relax. However, when we stood in line at the Boise airport to show our tickets and passports, I had been a bit rattled when the mother of one of the students rushed up to me. "You have to promise," she insisted, anxiously tugging at my sleeve. "You have to promise me that you will bring my son home alive. Promise me!" I promised.

That mother was in the forefront of my mind as we prepared for our first night walk into this rainforest.

Over after-dinner coffee and Milo chocolate milk drinks, I reminded the students to be alert for snakes. "Don't worry about the black-headed python and the carpet python," I told them.

"Just don't try to pick them up. But any other snakes? Shout a warning and run. And don't touch any big heart-shaped leaves with prominent veins, or you could get blisters. I don't think a cassowary would bother a group as large as ours, so that's OK. You're going to spend a lot of time looking up into the canopy," was my final warning, "but don't forget to look at what your feet are doing."

The students looked at me. *We get it*, their faces seemed to say. *Don't smother us.*

In the early darkness, we drove to the overgrown field and parked.

Tonight, we would be looking for possums. Three species I particularly wanted to see. The first was a chocolate-colored species with a white underside, the Herbert River possum, found only in a very restricted area. The other was the green possum, whose light-brown fur has a greenish tint due to the presence of algae on the hairs. In this area, I didn't think that fairly close to civilization, we would see the striped possum, a striking black and white fellow with long-fingered paws for digging under bark

for insects. We could look for him another night, deeper into the rainforest.

We began to walk along the lumpy path at the edge of the forest, shining our flashlights into the tree canopy.

Eyes stared back at us from a high branch. This was the common ringtail possum, one we had seen several times before. Ruby eyes of gray tree geckos met our flashlight beams, and moths danced around us. Then, there it was! Lying on a branch not far above us was a fat Herbert River possum. The long tail, curled at the tip, hung down, and we could see the patch of hairless skin on the underside near the end—the gripping patch Herbie would use for holding onto branches.

We met no snakes. We encountered no stinging trees.

Then, the jackpot: a striped possum on the trunk of a tree, with his long-fingered front paws stretched wide, looking like a slim, short-furred skunk. This was very fine. Now if we could just find a green possum . . .

From somewhere ahead in the trees, we heard a long moan, a strange, Halloween-ish sound.

That's odd, I thought. *Maybe it's a cow from that farm a mile down the road.*

We walked on. The moans became louder and more frequent. I noticed that the two students who were trailing the group had hurried to catch up.

The moan became an ungodly, roar-like, monstrous, repeated *UUUUUUUHHHHHHOOOO.*

"Maybe we should go back now," one of the students suggested.

I was wracking my brain. Was there some enormous, dangerous creature in the rainforest that I had failed to research? Or could a tiger, perhaps, have escaped from a zoo or been shipwrecked on the coast forty miles to the east? Or had a huge crocodile thrashed his way into this forest from one of the coastal rivers?

UUUUUUHHHHHHHOOO. Could there be dinosaurs here, hidden through the ages in the vast rainforest?

Then there was a crash nearby, as if a heavy bundle had fallen to the forest floor from a great height. The crash was followed by a diminishing series of thumps.

Should we go back?

I had to find out what was going on.

Then came another *crash*, followed by more thumps. *UUUUUUHHHHHOOOO*.

I sent my flashlight beam into the nearby trees, found a pair of large eyes, a brown-furred body, and I remembered. "Of course," I told the students. "The crashes are tree kangaroos. We're disturbing them, and they are bailing from the trees and hopping away into the forest." I shone my light into the animal's lovely brown face so that everyone could see.

UUUUHHHHOOOO.

But the tree kangaroo was not making the strange monster sound.

By this time, we were very close to whatever it was. *CRASH!* From fifteen feet up in the tree, the tree kangaroo leaped from its perch and hopped itself into the darkness.

The monster continued to boom out its moan.

It seemed to be higher up in that same tree. The students hung back, except for the fellow whose mom had made me promise to bring him back. We shone our lights up and up. The monster roared.

I saw a pointed white string hanging from a narrow branch and followed it with my light.

At the end of the string, which was a tail, sat a ball of gray-brown fur with two rounded ears. The creature was the size of a cottontail rabbit—*much too small to be making such a huge, threatening sound*, I thought.

The creature turned to face my light.

It was a rat, and it opened its whiskery muzzle and bellowed *UUUUUHHHHOOOOOOO*, the exact sound that the scariest Halloween monster in the world would make just before it ate you. I nearly dropped my flashlight.

"You are the giant white-tailed rat," I said aloud.

"*UUUUUHHHHHHOOO*," the rat replied.

"Nothing I read told me about their call," I said to the student beside me. Later, much later, I learned that October is the beginning of the mating season for giant white-tailed rats, and that they have a varied diet, consisting of fruits, insects, smaller vertebrates, fungi, and a sampling of electrical wiring—but they don't eat people.

"Let's go back to camp," I said to the students. "I think I've had enough excitement for one night."

The Ghost of Ravenshoe Rainforest

But the Ravenshoe rainforest had even more in store for us. Ravenshoe is a small lumber town, now quietly dying, set in cleared fields in the midst of a rainforest growing on hills, small plateaus, and cliffs of black lava. In decades past, some areas of the forest were logged, but the clearings were soon swallowed up by a jungly second growth. Foresters have kept some of the rides, or cut-in roads, open. Other rides filled in with trees and vines within a handful of years.

We had a permit from the Queensland government to live trap ground animals, to salvage roadkilled creatures, and to mist net for aerial creatures. One of our targets was a small marsupial mouse, the Atherton antechinus, a species that had been discovered only the year before. Very few specimens of the Atherton antechinus had been collected and the mammologist at the Queensland Museum, who had discovered the species, needed more.

"Marsupial mouse" sounds like a tiny, timid creature. However, marsupial mice are not mice as we think of mice. Yes, they are marsupials like wallabies, kangaroos, and bandicoots. They have pouches somewhat like those of kangaroos (only backward-opening). But some kinds of marsupial mice have even stranger qualities.

For one thing, there are times of year when no males are to be found. The rainforest floor is a kind of desert, where food is rather scarce because most of the energy of the ecosystem is at the tops of the trees in the sunlight. So, the Atherton antechinus lives in a world of limited resources, especially during the dry season. After the females have been bred and are carrying young in the pouch, all the males die, allowing precious resources to remain for the females and their developing young.

For several months, only females exist, carrying young that are attached to their pouch and cannot move from it for many weeks. Then, when the young come out of the pouches, there are males once again, baby males. Male and female babies grow up, mate, and the males die soon after. For someone used to the mammals of North America, this seems quite bizarre.

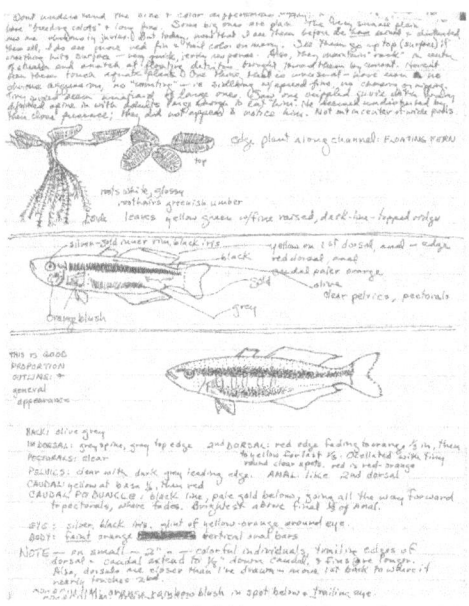

A page from my field journal from the 1984
C of I Australia Expedition.

Even stranger are the antechinus "mice" themselves. They are large, about the size of a very small chipmunk, and have long, pointed noses like the mice we know. BUT, and this is an important BUT—they have teeth almost like wolves!

Marsupial mice are not rodents, and don't have the two-teeth-up, two-teeth-down front chisel-teeth that mice and rats have. Marsupial mice have teeth on a smaller scale, but their teeth look like the teeth of large carnivores. And they can open their jaws into a huge gape, almost 180 degrees. A large rodent mouse might be able to open its mouth wide enough to bite your finger. An antechinus could take a chunk right out of your leg.

However, we were excited about the prospect of capturing the Atherton antechinus, teeth aside, and were very much hoping to catch a few males—because no one had seen a male Atherton antechinus. Ever.

We found an abandoned but still passable forest ride and set up lines of small box live traps in the forest out from this, as well as setting up mist nets for bats across it.

Each evening just before sunset we would bait the traps with dry oatmeal, which we would carry along in a cream-colored muslin bag. Then would sit quietly beside the weblike mist nets, large gossamer affairs spread across a whole ride and held up by very tall, segmented steel poles, hoping that bats would come in. Australia has no rabies, making it an ideal place where students can handle live bats and learn about them without the fear of getting sick.

The Ravenshoe rainforest is one of those forests that, to someone from North America, seems to have come from a storybook rather than from the ordinary geography of our world. The animals are strange, the plants are strange, and even the sounds are strange to one raised in the Northern Hemisphere well above the Tropic of Cancer. To this day, whenever I think of Ravenshoe, it feels like I am inside some enchanted tale, sliding down long vines of darkness into a dream.

We would sit silently on the damp ground a few meters away from the net poles as darkness fell. In the tropics, darkness *falls*—it does not creep.

There is nothing gradual about the coming of the dark in the tropics. One moment the sun is setting and ten minutes later, it is as dark as it will ever be that night. It takes some time to become accustomed to this phenomenon, especially for someone like me, a creature of the long northern twilights.

As darkness fell, the tree frogs, invisible to us, would call out. Fireflies, looking like tiny globes of pure gold, hung in the air at waist height during frog-calling time.

The frogs did not croak a few mild croaks. They were deafening. Some shrilled like police whistles or sirens. Others knocked like hammer-on-steel or hammer-on-barrel. Some shrieked, while others crackled or yipped. To carry on a conversation, we would have to cup our hands to the other person's ear and shout. Luckily for our sanity, the frog chorus would stop abruptly after about twenty minutes, and the fireflies would then turn off at once, all together, with perfect timing, as if someone had unplugged a string of lights.

Then, high above, we would hear the rustling.

I learned quickly that if I held a flashlight close to one eye, the outgoing beam would reveal the eyeshine of creatures touched by the light. We were soon possum-spotting in the undercanopy of the great trees high above.

The possums of the Atherton rainforest are strange, large-eyed furry creatures with small muzzles and prehensile tails, agile climbers that eat leaves and flowers every night high in the tree canopy. Some species of possum occur only in one small valley, while others have a wider range. Rainforest possums are beautiful and their eyes shine red-gold, like the planet Mars. Once in a while we'd see a baby clinging to its mother's back as she clambered among the branches.

We never tired of looking for eyeshine.

However, when we heard chittering and squealing, we switched off the flashlights and held very still.

Great squadrons of flying fox fruit bats would come in a twisting, turning flight down the dark forest rides, hundreds at a time, like the messengers of the Dark Lord in *The Lord of the Rings*. Flying foxes can have a wingspan as wide as four feet. Unlike the bats of North America, they have excellent eyesight, and all the time we mist netted for them, we caught only one. They also have strong jaws with efficient teeth. The one we caught chewed a three-foot hole in the top mist net and escaped in seconds, before we could reach him.

Once in a while, a chilling sound would come drifting through the forest, soft and directionless as mist—a low, breathy sound like someone blowing across the neck of a bottle: *ahhhhhh, ahhhhh, ahhhhh.*

When we heard this, the small hairs would prickle at the backs of our necks and down our arms. This was the voice of a cassowary. Cassowaries are large flightless birds weighing as much as a person and standing nearly as tall. Fearless, they can disembowel a human with one swipe of a clawed foot. We would stop speaking and freeze into immobility whenever we heard a cassowary, but we never saw one in the Ravenshoe forest.

We trapped each night for the Atherton antechinus but were unsuccessful. We captured some interesting bush rats and a few other little things (including a forest dragon lizard), but the antechinus eluded us. A little discouraged, we moved the traps away from the open ride and instead laid the line of traps, far into the rainforest proper—but still, the traps came up empty of antechinus each morning: no marsupial mice.

Finally, there was only one night remaining of our stay in the Ravenshoe rainforest. This was our last chance to capture the Atherton antechinus, as we had to break camp and begin the drive south in the morning.

We decided to move the trap line one last time, further south toward the Millstream River gorge, and since most of the students were engaged in packing up the camp, I volunteered to go set traps with Carrie, whose special project this was.

One of the fellows drove us to the new place on the road from which we were going to walk into the forest to set out the traps. He dropped off Carrie and me just before sunset with our box of traps and muslin bag of oatmeal bait. We knew we'd have to be cautious here, since not far away a 200-foot lava cliff bisected the forest, a precipice above the channel of the river.

Carrie shouldered the box of traps, and I picked up the muslin bag of dry oatmeal. We stood on the road's edge for a moment as we watched the taillights of the expedition's rented Toyota disappearing around a curve in the road.

Simultaneously Carrie and I looked at each other in dismay. We had left our flashlights in the vehicle!

"Come on, Carrie," I said, putting a bright face on it. "Let's hurry. We just won't go very far into the forest."

"OK," Carrie said stoutly. "We can hear the river if we get too close to the cliff. I think."

We stepped from the road into the forest, walking quickly. Frog-and-firefly time ended shortly after we entered the gloom under the trees. Vines clutched at our legs, and we stumbled over great, thick roots. We batted at branches with thorns and pushed onward, hurrying.

Carrie would take out a box trap, set the treadle, hold it out to me, and I would throw in a handful of dry oats from my cloth bag. Then I would place the set trap on the ground. Then we would take twenty steps and place another.

We were going very fast, trying to beat the dying of the light, but, of course, we failed. After the nineteenth trap, we looked at each other in dismay: the light was virtually gone.

"Let's turn around and go back," I said. "We'll set the twentieth trap and turn around. We can move ten steps to one side and set a parallel line of traps on the way back, instead of doing one long line straight into the forest."

"Yes, let's!" came the eager reply. I took eight more steps. Suddenly the deep, hollow sigh of a cassowary filled the air.

Ahhhhh, ahhhhh, ahhhhhhh.

I hesitated—and felt a breath of cold dampness on one cheek. Then I heard the rushing of water.

"Stop, Carrie!" I said softly, for one does not shout when a

cassowary is nearby. "Stop in your tracks. We're at the edge of the cliff."

We were only an arm's length from the edge of the drop to the river.

Very slowly and carefully, we backed away from the invisible gorge. A moment later, I smiled. This told me exactly where we were, and how to direct our trap line back toward the road.

Walking quickly lest we lose our orientation in the darkness, we counted steps. Carrie held out traps. I reached into my bag and threw bait into each one, then set it on the ground, and we moved on. We were not lost. It was dark, but we were doing the trap line properly, and we were going to be fine.

Carrie held out the final trap, I baited it, put it down, and we were done.

Now we could simply walk the remaining fifty yards or so to the road and wait to be picked up. We talked quietly, anticipating a hot dinner and comforting cup of hot chocolate beside the fire back at camp. We moved through the forest in single file, with Carrie going ahead. I held the muslin bag in front of me. It was the only thing I could see, the faintest possible pale blur.

The bag was rising into the black air, but I was not lifting it.

This was not logical.

I could hear Carrie just ahead and did not want to fall behind, so I kept walking. The bag rose as I walked, now up to shoulder level. From the darkness behind me came the hollow, ragged sigh of a cassowary.

Ahhhhhhhh. Ahhhhhhhh. I shivered.

This cannot be happening, I thought. This was one of the few moments in my life when I felt like screaming, but one does not scream when a cassowary is near. I looked at the bag and took another step.

The bag was now at chin level. I was farther behind now; the faint sound of Carrie's steps was fading.

Grasping the bag with both hands, I strode strongly forward. The bag lifted above my head. I could feel it pulling away as it rose up and up.

"Ghost?" I whispered crazily. "Ghost, give me my bag!" I could not hear Carrie anymore. I pulled on the bag with all my strength and could not pull it down again.

"Stop!" I said louder than I had intended. "You are not going to take this bag!"

Still holding the bag in both fists, I took a giant step forward and found myself swinging in the air as the bag rose into the black canopy of the forest above me. This was downright frightening.

But things must be logical—mustn't they? And anyway, this was our only bait bag, and I was not about to lose it to a ghost.

Scrambling and kicking, I swung myself backward. The bag came down a few inches. Stopping stock-still on my tiptoes, I stood in the blackness while my brain ticked over a little. I took two more steps backward and the bag came down to my chin. Without letting go, I retraced my steps into the forest, and soon the bag was at my waist.

I stepped forward, and the bag rose into the air. "Go away, ghost!" I said.

The cassowary answered, closer now. *Ahhhhhhhh.*

I wrenched at the bag, heard it rip, and I ran forward, stumbling. I continued to run, banging into tree trunks, and lurching over tangled roots.

And at last, there was the ribbon of road, and a car.

Light from yellow headlamps silhouetted Carrie, who was shouting my name. I almost fell from the forest onto the road shoulder. I had beaten the ghost, because I still had my bag!

Carrie and I scrambled into the car, shut the door against the night, and strapped ourselves in. Brad, who had come to get us, turned smoothly around on the road, and our adventure was over.

I found my flashlight on the seat and clicked it on. I examined the muslin bag. Caught in a small rip near the neck of the bag was a thorny wand. Supple as a willow switch and slender as fishing line, it was lined with wicked, curved hooks.

I thought a bit. Yes, this was the ghost of the rainforest, a strong, whippy branch dangling from a tree, that reached invisibly down and snagged—and held—my bag! The farther away I marched, the more it had pulled at the bag.

The next morning as we were packing up the final bits of camp, I showed the thorny branch to the campground owner and asked him if he knew what it was. "Oh, that's a lawyer," was his startling reply.

"What do you mean?" I asked. "A lawyer?"

"Yes," he replied. "That there is the lawyer vine. It's a kind of a palm tree vine that grows on some of the trees and hangs down these cat-claw branches. They say that once a lawyer has got you, you cannot get away!"

And did we trap an Atherton antechinus during that last, strange night in the Ravenshoe rainforest? Oh, yes. The next morning, we found a large female caught in the farthest trap, the one near the edge of the cliff. She bit me and left quite an interesting impression of her teeth on my leg.

Confessions of a Skunk Trapper

Why would anyone want to trap skunks? Why would anyone be paid to trap skunks? No one in their right mind, that's for sure. Nevertheless, I once trapped skunks nearly every day, and liked it.

In the summers from 1979 to 1986, I worked for the University of Idaho on the original research project that defined the boundaries of the then-proposed Snake River Birds of Prey National Conservation Area.

Birds of prey need prey, and to help define the boundary of the proposed sanctuary, we set out many trap lines of live traps to see where prey were most and least abundant.

One spring, my three trap lines were rows of large wire-mesh box traps that I baited every late afternoon with apple and cabbage. One of my trap lines, Rim, stretched along the rim of the Snake River Canyon, atop black lava walls falling a sheer 200 feet to a steep, rocky slope that led down to the Snake River 200 feet farther below. The second trap line, Talus, I positioned in the canyon itself, on that steep slope below the cliffs. The third trap line, River, lay parallel to the Snake and very near the water, in and out of miniature meadows bounded by willow, currant, and sumac bushes.

Every morning that spring, I would rise before the sun and drive south on Swan Falls Road to begin my checks of Rim, Talus, and River.

Each little creature I caught was carefully weighed, tagged, recorded, and released. For each, I would record the species, weight, sex, and condition, let the little one hop or scurry away, and move along to the next trap. By noon I would be finished. After eating my lunch and writing up my notes, I would re-bait and re-set all the traps before returning to the office, only to drive back to the canyon early the following morning.

I captured cottontail rabbits, woodrats, ground squirrels, chipmunks, jackrabbits, porcupines, and marmots. After the first two days of running the trap lines I thought, *this is the world's best job. I'm out in the spring sunshine where I get to look for wildflowers, hear the meadowlarks singing, see the falcons flying over, watch the otters playing in the river—and I get to handle all these small creatures without harming them.*

On Rim, I often caught bushy-tailed woodrats, those lovely and gentle pack rats with the furry tails. When I released them from the weighing net, the young ones, gray as clouds, would often sit on my shoes and eat cabbage and apple bits from my hands.

At River, porcupines were in residence. One has to move slowly to release a porcupine from a live trap without setting them off. I was very glad I didn't have to weigh the porcupines. After all, porcupines would not be the usual prey of hawks, owls, and eagles. One particular male porcupine, #12, or George, was so avid for the cabbage-leaf bait that every morning I had to poke him with a stick to get him to leave the trap, and I had to shut down the trap's door quickly or he would waddle back inside at once. I caught sweet-natured George every morning in the same trap, and I found myself loading that trap with more and more bait, since he would head for the trap as soon as I began walking away after setting it in the late afternoon. George was spending so much time in the trap that I worried about him getting enough to eat.

Chipmunks and ground squirrels were pretty, though they would try to bite. Marmots were relaxed and docile, and cottontails were no trouble at all. I loved running the trap lines. It hardly seemed like work, except for the getting-up-early part.

Then I caught my first skunk.

It was on Talus, and it was a spotted skunk. Spotties are very small, extremely attractive skunks, sporting glossy black coats sprinkled with white bars and dots, and having plumy white tails. When they feel threatened, they stand on their "hands," stamp their front feet, fluff out their tails—and *squirt*! They let you have it from their scent glands, at high velocity. Spotties are quite irritable, and this one was stamping its front feet even before I got to its trap.

I knew that the little fellow would spray me right through the wire mesh if I got close enough to open the trap's door. What to do, what to do?

I went about recording and releasing all the other creatures caught on Talus that morning while I pondered what seemed to be the only choice—to take my medicine and get squirted. I couldn't walk away and leave the little skunk in the trap to die in the heat of the day. *Arrgghhh!*

The moment of truth arrived. I looked wildly around for help. Then I saw it!

Down by the river was a pile of ugly junk left by boaters and fishermen. In this pile lay a scruffy black rubber raft.

The raft was about seven feet by five, fully collapsed (it no doubt had been discarded because it leaked) and still had a long rope tied to a ring at the front. "Aha!" I said aloud and rushed down to the riverbank to claim it.

Holding the raft in front of my body, I approached Spottie's trap. When I was four feet away, he let loose with the spray. I raised my raft-shield with the speed of light.

I have a clear memory of the yellow-green droplets shining in the sun as they flashed through the air toward me—and how they splatted on the rubber raft as I ducked behind it.

The stench was horrific, eye-watering—but at least it was not *on me*. And my instructions were *not* to weigh skunks!

I flapped the raft over the top of the trap, completely covering it. Gingerly, I reached under the raft and propped open the trap's door with a rock. Then I took up the rope and backed away. Fifteen feet from the trap, I used the rope to pull the raft from the trap.

Spottie sprayed again, but I was out of range. My hand was a little stinky, and the raft was mighty stinky—but it had saved the day. When Spottie had hidden himself under a boulder, I tied the rope to the trap, folded the wet side of the raft around it, and carried my perfumed bundle down to the Snake River.

At the water's edge, I held the rope in the middle and threw both raft and trap into the current, anchoring the rope with a big rock. I then went downstream and checked the River trap line. When I came back for the raft and Spottie's trap to pull them from the river, they both smelled a bit skunky, but only a bit.

Back went the trap to Talus, re-baited for the next day. Into the back of the government pickup went the blessed raft.

I caught skunks about three times a week during the months I trapped Rim, Talus, and River that year—both the little Spotties and the larger striped skunks. Most of the striped skunks became so accustomed to being caught and released that they stopped spraying.

I would approach a skunk-filled trap cautiously, saying idiotically, "Nice little skunk. Nice little skunk." The recaptures would lower their heads and face the traps' doors, and I could open the doors without fear that the stripeys would spray me. But first-time-caught skunks would panic and try to claw their way out of the traps. I held up my raft when releasing those. Those skunks always sprayed. The spotties always sprayed too, and each time, my dear rubber raft saved my bacon. My raft got skunkier and skunkier as the weeks wore on.

Several times, I had to shut down the trap lines for a few days to do other research. On these days my work partner was a person whom I will call "Frank."During the summer field season, Frank was living in an old trailer on the property of a local rancher. When I worked with Frank, I would drive to the trailer and pick him up, head to the desert for the workday, and deposit him back at the trailer at quitting time.

Frank was *frank*. He was very frank. His mission in life seemed to be to convert me to his religion, which appeared to be austere and joyless. After an interval of politeness, I asked him to stop trying to convert me during working hours. He ignored my request and became even more frank. As the field season progressed, I grew weary of Frank. I much preferred the skunks and porcupines.

For some fun to mark the end of the field data-collecting season each year, we biologists and technicians would have an open-air potluck party down by the Snake. We would play volleyball, grill food over a fire, and eat ourselves silly. Frank's trailer was only a few hundred feet from the area chosen for the party, and several times we asked him to come—but he declined.

On the Friday evening before the Saturday party, I dropped Frank at the trailer. This time I got out of the pickup and again asked Frank to come to the party. I had the idea that some relaxation might help him become less rigid.

"Not a chance," Frank replied. "There will be beer. Someone might dance. There might be a radio with the *wrong music.*"

"Some people will bring beer, but you don't have to drink any—I don't," I said, trying to be friendly. It was difficult. I was hoping he would invite me into the trailer for something cold to drink, so we could talk.

"No," Frank said. "This kind of party is from the devil." He shut himself into the trailer. I turned to the pickup, and my eyes found the faithful rubber raft lying in the pickup bed, steaming skunkily in the hundred-degree heat, even folded as it was with the smelly part inside. The field season was now over, the traps cleaned and stored. What would I do with the raft? I couldn't take it home—the thing would stink up the whole street.

Suddenly I noticed a dark gap in the aluminum flashing that served as a skirt for the old trailer.

Saturday was pleasant and very hot. The party was a great success, attended by dozens of researchers, spouses, children, and dogs. After an afternoon of volleyball, potluck dishes were set out and we lit a fire to grill chicken, hamburgers, sausages, and steaks.

Two fellows walked over to Frank's trailer for a last attempt at inviting him. When I saw them trudging back, I could tell from their dejected looks that Frank had turned them down. "Wow," Jim remarked, "it sure stinks over there. Must be a whole family of skunks living under that trailer."

"Yeah," said Rich. "You know that trailer has all kinds of holes in the floor where the stench can come through. I don't know how Frank stands it in there!"

A faint odor of skunk clung to Jim and Rich as they sat beside me on the grass. It turned out to be a beautiful, starry evening down by the Snake River.

What Not to Wear

For a couple of years in the mid to late 1980s, I was the Snake River Birds of Prey Natural Conservation Area's habitat biologist (that is, I worked with the plants and small mammals).

Morley Nelson, may he rest in peace, worked tirelessly for many years on behalf of raptorial birds in Idaho: hawks, falcons, owls, and eagles. Morley was an avid falconer, and his best love was the eagles. He flew several trained eagles for many years.

Morley was instrumental in bringing about two major conservation triumphs for raptors. The first was to convince the state power utility, Idaho Power, to design and construct all new power lines and towers in such a way that eagles, owls, and large hawks could no longer span two lines with their wings and electrocute themselves. Morley also convinced Idaho Power to begin a major effort to replace all the old, raptor-deadly power lines and towers with the new, raptor-friendly ones.

Morley's second achievement, with the help of many others including Cecil Andrus, Secretary of the Interior, was the establishment of the Snake River Birds of Prey National Conservation Area, an area of grazed public land shrub-steppe that was to be protected from development and managed for healthy populations of raptors and their prey. I had been proud to work on the initial surveys of prey animals and their habitats to establish the boundaries of this 1,400,000-acre preserve, which boasts one of the largest populations of nesting raptors in the world.

Several years after the establishment of this natural area, it was re-named the Morley Nelson Snake River Birds of Prey

National Conservation Area—quite a mouthful! (Try putting that into a column in a graph.)

On this occasion, some organization, I now forget which, organized a banquet with speeches. The guest of honor and main speaker, of course, was Morley Nelson. I was invited to attend, and I knew from my boss that I was expected to make an appearance.

The banquet was a buffet held in a large conference room. White tablecloths and flower arrangements covered a large, square formation of tables that nearly filled the room. The front table was reserved for Morley and the other speakers, while the rest of us headed to the buffet line and then went looking for seats that had not yet been taken.

About the time I got to the serving table, the buffet ran out of food. My portion was four small pieces of pineapple from the dregs of a fruit salad, and four green beans. *Oh, well,* I thought. *This is Morley's night, no biggie on the food. I can grab a burger on the way home.*

As I searched for an unoccupied chair, I was pleased to run into another biologist, Bill Clark, with his wife Mary, longtime friends I had known since college. Bill's plate had about the same amount of food as I had been given; Mary had a bit more. We laughed and moved toward the front of the room, hoping for seats close to the head table.

To one side of the head table, a five foot tall wooden stand held a golden eagle, a mature female, her two-inch claws grasping a wooden crossbar tightly enough to dig into the wood. The huge yellow feet were clenching and unclenching on the bar. I noticed that she had about six feet of cord anchoring one of her leather leg jesses to the stand.

I said to Bill, "Even though she's hooded, that eagle looks stressed to me. I think we should find a seat a little farther from the head table." Bill nodded and we sat about fifteen feet away from the eagle's stand.

I savored my bits of food while several of the dignitaries spoke. I sighed; I was hungry. To distract myself, I looked around at the other diners.

Many of them were local biologists and raptor biologists of

national reputation, whom I knew well. Most of these folks were dressed similar to me as well: jeans or khaki slacks, cotton button-down shirts or turtlenecks, with a corduroy or twill sport jacket here and there. But near the head table sat someone who stood out, a woman carefully coiffed and wearing sparkling jewelry.

She wore a beautifully made floor-length gown of mauve crochet-work, lacy and feminine. The dress was formal and elegant, with a neckline high in front and dipping very low in back, finished with a gracefully draped roll of crochet just above the waist.

Holy cats, I remember thinking as I ate one of my green beans. *It would never have occurred to me to wear a formal gown to this event.* With glazed eyes, I stared at Ms. Mauve's bare back as the speeches droned on, wondering if she were cold. The air conditioning was rather fierce, and I was glad I wore long sleeves.

After what seemed forever, Morley got up to speak. As he prepared to take the stage, he removed the hood from the eagle's head. I'd heard Morley speak a number of times and knew that he often had an eagle on his arm as he spoke from a stage or podium. Morley bent to unfasten the cord clipped to the eagle's jesses. And then, yes—all hell broke loose.

Someone bounced up and took a photograph—with the flash on—of Morley and the eagle.

The eagle took offense, instantly. Dodging Morley's grasp, she launched herself straight—you have guessed it—at the beautiful bare back of the woman in the crocheted gown.

There was much flapping and screaming, some of which was done by the eagle herself. The wooden stand followed the eagle and smacked across the backs of several people at Ms. Mauve's table.

Chairs toppled as people pushed out from the tables, either to run or to help capture the rogue eagle. Bill, Mary, and I got up, and began to back away from the melee.

A flower arrangement smacked me in the knee before crashing to the floor and shattering the vase. Plates and food flew everywhere. Trying desperately to flee the eagle, Ms. Mauve had latched onto the tablecloth with both hands, and as she yanked and scrambled, everything on that table slid, fell, or flew through the air.

By this time, the eagle, just as panicked as Ms. Mauve, had finally found something to grasp. She sank her talons into the lovely, draped roll of crocheted fabric at the back of the mauve gown and held on for dear life, all the while beating her powerful wings against Ms. Mauve's back and head—battering the woman, the people trying to help, the table, and Morley himself.

If you have never seen an eagle close up, you may not realize just how strong those big feet are. The foot tendons have a strange rachet of little nubs that allow the eagle to hold its grip for long periods of time, and the grip of the average eagle has been estimated to be 400 pounds per square inch. The claws, too, are formidable. Once, when watching another biologist attach a satellite transponder to another golden eagle, I saw the eagle slide a claw, seemingly with little effort, all the way through the palm of the biologist's hand. He grinned, slid the claw out, and showed me old scars on both hands from eagle claws.

Sliding a claw out is one thing, but getting a mature, panicked golden eagle, to release her grip—that is something else.

By this time, two men had corralled the eagle's wings by holding the bird in a bear hug, one man on each side of Ms. Mauve, who was still shrieking. Morley and others were working on the feet.

I looked at Bill and mouthed a silent, "See you." True to the long-time motto of my family, I was *First to Leave.*

Just before I slipped out the door, I heard a man's voice saying, "I think we're going to have to cut the dress." And a woman's voice cried, "No, don't cut my dress!"

I don't know how the evening ended, or if the dress made it, but I was ready for a hamburger.

Fishing with a Fur Coat

Have you ever hit rock bottom? It's never a good place to be, especially when you're over 40, when you have been cut loose from your marriage of 20 years and replaced by someone else—when you are unemployed, having quit your job to go on that last long field expedition, when your mother has just died, you have lost your home—and your heart-dog, Duncan, partner in so many of life's adventures, has gone beyond that mythical bridge.

Late that summer, I had taken the first wobbly step out of the darkness.

I had landed a job, as copy editor for a magazine publishing company in Boise. Some of the people there were wonderful, and some were not. Publishing seemed strange to me after my years in field biology. My supervisor made more than twelve times my own salary. Mostly, he conducted meetings. But now I could pay my bills and had no trouble doing my own work.

On a Friday evening in late September, I drove alone to my family's cabin in the Sawtooth Valley. That summer, I hadn't been able to afford the gasoline to go fishing, but now I could. I was in need of a time to be wild, a time of doing something that would occupy my mind and drag it away from whirling around the drain of life, something beautiful.

I arrived at the cabin at midnight, tossed my sleeping bag onto one of the beds, and fell asleep, not even bothering to open the blinds and shutters.

***

A thin line of burning sunshine fell across my face from a gap in the blinds, and I sat up. Morning had come, and I had made it to the cabin!

I opened the front door, stood on the small deck and took in a deep breath of cold, pine-scented air. Vat Creek rushed happily downhill a few yards to the south. From somewhere high in Walter, the big Douglas-fir next to the deck, a squirrel scolded me. The snowberry bush at Walter's base had already turned autumn-gold, and the sky arced blue and cloudless, a bluebird morning.

I walked around the cabin, opening the shutters. I remembered that I had brought bacon, eggs, and coffee: time for breakfast.

After breakfast, armed with my fly rod, two cans of Diet Coke, and a scrambled egg and bacon sandwich, I climbed into my old blue Bronco II, Little Falls. Trailing a plume of dust, I headed down into the valley to find a place to fish.

I needed a place that wasn't easy, a place where flyfishing would be difficult enough to occupy my mind and keep the black thoughts away. Therefore, the Salmon River was out. Valley Creek was out. Frenchman Creek was out. Even Elk Creek was out, and Basin Creek was farther than I wanted to drive. Fisher Creek! I thought then. *I'll go to Fisher Creek!*

Fisher Creek was only a few minutes' drive from the cabin. One of the meadows at the edge of the pines sported some cabins; but farther upstream, where I was headed, there would be nobody, and that's what I wanted on this bright September day. Nobody.

In this area, the creek wandered through what had once been a swamp. The ground there was uneven and damp, a mess to fish, with a maze of willows bristling with branches that could snag a fly line anywhere a flyfisher chose to stand and cast.

But there were cutties in there, cutthroat trout, and lots of them, because any fisherman with sense wouldn't even attempt to fish inside that tangle. It would take skill and concentration, and maybe a few cusswords, to pull cutties out of that part of Fisher Creek—just the challenge I needed.

I pulled Little Falls off the dirt road and set up my rod. Back at the cabin, there hung on the kitchen wall a cast iron skillet, and there was butter waiting in the fridge. Time to catch my dinner.

I shoved my lunch into the old canvas creel and fought my way through mud and willows to the hard gravel of the creek bank.

The first trout I caught were little fellows, fingerlings. I was happy to let them go, and concentrated for hours, laying my fly lightly on the water in exactly the right places without getting hung up on the cast or the backcast. This was a challenge, all right.

I forgot about time and circumstances and all manner of change and evil. The world at 7,000 feet was clear water, shining gravel, a dipper bobbing along the banks, billowing grasses, tiny wildflowers, chattering squirrels, and my favorite of all the perfumes in the world, that clean-water, wet-willow scent that reminded me of going flyfishing with Dad and Mom—the very best times of my life.

Finally, however, I noticed that the shadows of the pines on both sides of the willow-swamp were growing longer. To the west, the sun was sinking over the high, serrate ridge of the Sawtooth Mountains. It was time to find a larger hole or two and catch some skillet-sized fish, four for dinner, two for breakfast.

I fought my way through willows and a jackstraw of half-drowned, bleached pine trunks, to a long, deep hole with a wide gravel bank and a decent space behind for casting. Time to get serious.

I got serious.

The sun disappeared behind the high ridge. I stood in the stream just above the hole, casting. In no time, I caught four pan-sized fish. Two more fish and I could drive home in the early twilight to cook dinner and read myself to sleep.

I got a bite from a big one, but he slipped away. If I could just catch him, he would be enough for breakfast all by himself.

A dark head lifted from the water near where I had placed my fly. Narrow jaws held a small trout. The creature swam to the bank, deposited the three-inch fish on the gravel, and shook itself all over.

It sat on its haunches and stared at me.

"You're a mink," I said softly. "Hi. I'm a fisher."

The mink ducked its head and slid back into the water. I got another nibble from the big trout, but he slipped away again.

The mink surfaced with another small fish and deposited its catch next to the first.

I caught one. Too small. I threw it back.

The mink caught one. Just right for a mink. Laid it in line with the other two.

It's a contest! I thought. I caught another little fellow and tossed him back. The mink laid another fish in line with the others on the gravel.

He caught one. I caught one. He caught one. I caught one. Light was fading.

Then I hooked the big one and fought to reel him in. He was a keeper. Breakfast.

Brook trout and fly rod, Stanley Basin, Idaho.

The mink picked up the largest of his fish, bobbed his head at me, and disappeared into the willow tangle. "Who won?" I called after him. Laughing a little, I made my way back to Little Falls and drove down Fisher Creek Road to the highway.

I'll do this again, I thought. *On Monday, when people at work talk about what they did over the weekend, I'll have a story, too.* I turned up the dirt road to the cabin. Two mule deer at the side of the road moved away from Little Falls' headlight beams.

"This is my life now. It might not be so bad," I told myself. "Maybe there are more best times to come. One of the assistant editors seems really bright. Scott said he would like to learn to flyfish." Back in my days of guiding the teenage children of Dad's clients, I had taught many people to flyfish.

I can teach Scott, I thought. *He seems nice.*

Today is our 35th wedding anniversary.

Four Fools

Divorce does strange things to people, especially a divorce after a long marriage, a marriage that ended sadly. I was coping, but it wasn't easy.

After the initial depression, I felt that I had to change my attitude. "I can do it, no matter what it is," became my new, rather ridiculous, motto. "I don't need anyone else. I can do anything and everything on my own." I was doing all right in my new job as copy editor for a magazine publishing company. Since many of my old friends had suddenly, and voluntarily, disappeared from my life, I made new friends (which has never been easy for me). I tried new things, ventured into new places.

Some of my friends were students at the College of Idaho. And one of these students wanted to do a senior research project on the physiology of the kidney of kangaroo rats. The k-rat kidney is a marvel, so efficient at removing water from dry seeds that the rats produce urine not as a liquid, but as a pellet of solids. My friend wanted to keep a few k-rats in the lab for a few months and by weighing their food and excrement, determine the amounts of water, calories, and nutrients extracted from their food. Then he'd release the rats back into the desert. But he had tried and tried to catch some k-rats, obviously not trained by the fun-loving crew who had trained me, and hadn't been able to catch any.

I was an experienced k-rat catcher and was very fond of these appealing little fellows with their tan and white coloring, huge bright eyes, long furry tails, tiny front legs, and upright hopping habit.

When I was a freshman at the C of I 25 years before, I had learned from older students how to conduct a successful k-rat hunt. Equipment needed: a flashlight, a butterfly net, a good sense of direction, and a vehicle capable of handling very bad desert roads in the dark. A bunch of us students would pile into someone's old pickup, drive out to the desert, and watch for k-rats to go bouncing across the little roads. I say "roads," but most of these could barely even be called two-track trails. The k-rat roads were rough and rutted, full of big rocks and dust traps. Sometimes we'd get stuck, but a group of students all together can push a pickup out of almost anything.

Once we got into the deep desert, two people with nets would ride the fenders. Others would sit in the truck bed. Our truck would crawl along at five miles an hour, and we'd start looking. When the truck's headlamps showed a pale rodent hopping across the track, we'd train our flashlights on him, jump off the truck hood or out of the pickup bed and, butterfly nets fluttering like white moths in the beams from headlights and flashlights, would go after him.

We learned that a person could often outrun a k-rat through short grasses and desert shrubs—but if the k-rat got to his burrow, we'd be empty-netted.

K-rat hunting was fun. When we'd catch one, we'd look at the little fellow closely, pet him, and then, most of the time, would let him go.

For two years of my undergraduate life, I had a k-rat (forbidden item) as a pet in my dorm room. Cleopatra was her name. Cleo was given to me by my friend Terry Uhlman. She lived on dry seeds, and since she never produced liquid pee and her droppings were odorless, she was a fine and gentle pet. She'd hop across my bed to nestle in my lap or climb my sleeve to sit on my shoulder as I studied.

Since k-rats are nocturnal, Cleo was content to sleep in a shoebox in my closet during the daytime. She became active only after sundown when I was usually in my dorm room studying. Cleo didn't bark or howl, she didn't shed, she didn't chew anything but her seeds and the paper towels I put in her box, and she was small—the perfect pet for a student who often pulled all-nighter study sessions.

Twenty years later, in a full not-quite-sane "I can do anything" mode, I was ready to go k-rat hunting by myself, in the deep desert—in late November, no less—to get a couple of k-rats for John and his physiology study.

I brought a pair of jeans, a t shirt, a sweatshirt, and some old shoes to work, and as soon as everyone else had left the building, I changed out of my work attire and climbed into Little Falls, thankful for his four-wheel drive. The net and flashlight, plus two little carrying boxes, I had already stowed in the back. I had also brought my pistol because you never knew.

I headed for the deep desert. First, I drove through the small farm town of Kuna. On a weeknight, Main Street was quiet. I didn't even see a fight on the sidewalk outside the Red Eye Saloon. I turned south.

South on Swan Falls Road I drove. After a few miles, I had left the last farmhouse and had entered the million-acre Snake River Birds of Prey National Conservation Area, where I had worked gathering data for several years. Though I was familiar with the little roads in the area where I planned to go, I had never been in the Birds of Prey Area at night.

This was the deep desert, many miles from the nearest building. Even now, cell phones and military radios don't work well in what we call "the hole." But the hole does have a satisfactory population of kangaroo rats.

I hoped to catch the rats before moonrise, since (sensibly enough), fewer k-rats would venture out in the light of the moon. Instead, the little hoppers would try to foil the hunting owls by scurrying about for seeds to fill their fur-lined cheek pouches before moonrise, in the darkest of the darkness.

Just as Swan Falls Road turned sharply west before its descent into the Snake River Canyon, I turned east onto Moore's Road, and then south onto a smaller track and soon found myself in the lumpy flat just to the west of Big Foot Butte. I rolled down the driver's side window—*brrr*—and put my net at the ready.

There! Hopping through the silvery winterfat shrubs was the night's first k-rat. I stopped the Bronco, grabbed the net and flashlight, switched it on, opened the door—and the k-rat was nowhere to be seen. "Well, at least I'm in a good place," I told myself. "There will be more." I trudged back to Little Falls, empty-handed.

A fat badger, trundling heavily down the road, shuffled to one side and showed me his teeth as I passed, and after half a mile, I had seen nothing more than a handful of pocket mice and deer mice.

A hundred yards farther on the two-rutted track, another k-rat hopped into view. I stopped the Bronco, grabbed the net and flashlight, opened the door—and again, the k-rat had vanished. *Hmm,* I thought. I jammed the flashlight, still on, between my knees.

Another k-rat danced lightly across the ruts. I stopped the Bronco, grabbed the net and flashlight, opened the door—and you guessed it—the rat was gone. "THING!" I shouted to the sky, my harshest expletive. I had to think. It was taking me too long to get after the k-rats. I'd never capture any at this rate.

A mile or so to the south, a family of coyotes howled, their ice-needle cries spiking up and down the scale. Above arced the Milky Way, brilliant with clouds of stars so far from any man-made lights. The night felt like an enormous cape of cold black velvet, swirling widely about, black as black; comforting, familiar, and obscuring everything—mountains, buttes, and canyons. I loved this place.

I took some deep breaths and exhaled, thinking. The narrow flashlight beam caught my breath curling in a white mist around my face and flowing away in the freezing wind.

"OK," I said to myself, marching back to the Bronco. "I'll keep the flashlight on, between my knees. I'll put the net handle vertical in the door to keep it open. I'll stick out my left arm, and hold the door shut with my elbow. Then when I stop, the door will be open, and the net will be ready." This had to work.

I knew from the glow on the eastern horizon that precious little time remained before the moon would rise over the Danskin Mountains and flood the winterfat stand with silvery light. The kangaroo rats would dive into their burrows, and the night's hunt would be over.

Movement. There was a rat! I turned off the Bronco, jumped

out, leaving the door open, chased him ten feet through the winterfat—and got him! The rat was a fine young male. Into a holding box he went. I could hear the rattle of dry wheat as he hopped around in the box. But I knew k-rats. He wouldn't be very distressed, and in a few minutes, he would be picking up the wheat and stuffing his cheek pouches full.

This was working!

I drove further south, into the heart of the "hole." I slowed. I knew this area. The soil here was as fine as talcum powder and deep, seemingly bottomless.

The National Guard trained here, and after a few passes with any vehicle, the ruts would look shallow but could be deep, filled with fluffy dust two—even three feet deep. What looked like a 3-inch-deep rut would be a three-foot-deep rut, and a vehicle, even one with excellent clearance like the Bronco II, could find itself high-centered, wheels spinning without purchase in the dust—and it would be stuck, seriously stuck.

I drove off the track, driving around suspicious-looking ruts, and I put the Bronco into 4x4 low. If I hit one of those dust blowouts, I'd never get to work the next day. Hardly anyone, even Guard soldiers, came this far out into the desert on a weekday in November. If I got stuck, I'd have to walk fifteen miles out to a real road and hope for a ride.

"Nope," I told myself. "I can do this. I've driven these tracks many times before, in the daytime. But one more rat and I'm heading for home."

Five minutes later, I got that rat and dropped him into the other box. The Bronco's clock showed midnight. Then the moon crested the black ridge of the Danskin Mountains. I had netted that second rat just in time.

Stark moonlight lit the gray winterfat, making ragged shadows of the small bushes. The slanted bulk of Dorsey Butte showed straight ahead, and to the northwest loomed Big Foot Butte, huge and domed. The black featurelessness had lifted, and the landmarks I knew so well were back.

But, wait. What was that odd, blocky form just ahead? Was it a cow? The ranchers had rounded up the cattle many weeks ago, just as they did every year. The cattle wouldn't be back in the desert for nearly another month. The ranchers had taken the water tanks back to their ranches, too. Could the thing be a deer or an antelope, or even a wild horse?

Then the moonlight showed me a glint of metal, and then a plume of dust puffed to one side of the blocky object. And there came a shout.

"Hey!"

Part of that strange object was a person.

I scrabbled for the pistol under the driver's seat and jammed it into the waistband of my jeans, in front, where it could be easily grabbed. Then I pulled my sweatshirt down over the gun and got out of the Bronco, slipping the keys into my pocket.

Someone was running toward me. I nailed him with the flashlight. "Stop!" I shouted.

He slid to a standstill ten feet away, dust billowing around his feet. The flashlight told me he was a young man in jeans and boots, naked to the waist, a black man. I took a closer look. He wasn't black. He was covered in dust and sweat, and the sweat had turned the dust to a dark film on his skin. His eyes stared at me palely from his darkened face.

He turned to shout over his shoulder, "Hey guys," he said. "It's a woman!"

"What the hell?" A faint voice came from somewhere down the track.

"Please, please, we need help," the young man said.

"Let's go to where your friends are," I told him. "You will not come any closer, all right?"

"Sure, sure, fine," he said. "Come on. We're stuck."

I followed the man fifty yards down the dust-rutted road to the odd object I had seen in the moonlight.

A long, low car—an old muscle car from the 60s, had bottomed out in a dust blowout and was high-centered there. Two more mud-and-dust young men heaved themselves out of the rut. One held a bent hubcap. He had been digging with it.

They surged toward me. "Stay there," I said, "and I'll help you. But don't come any closer."

"Whatever you want, ma'am," one said. "Whatever you want. Please."

"I'm out catching kangaroo rats," I told them, vastly relieved that they didn't seem violent. "What are you doing out here?"

"We work at a garage in Meridian," one told me, dragging a toe in the dirt. "We had the day off, so we got some beer and thought we'd come out here and ride around. Please take us to Meridian. Please! We'll all get fired if we aren't at work at 8 o'clock sharp. We've already had some warnings. Mr. Murphy is really strict."

"My dad is going to kill me when he sees his car," one of them, the smallest of the three, muttered. "He's been restoring this one for two years. We shouldn't have taken it."

"Shut up, Carl," the first young man, the tallest, said. "We've got to leave the car here. No choice. We ran the battery down. It's out of juice and it's out of gas, too."

The second fellow continued, "We drove around out here all afternoon, drank our beer, and then kind of took a little nap. When we woke up it was dark, and we got lost—and then we got stuck. And that was . . ." he checked his watch, " . . . five hours ago. We've been digging for five hours."

"My dad is going to kill me, kill me," the smallest one repeated.

I stood ten yards from them and assessed the situation, the hidden pistol a comforting pressure against my stomach. "OK," I said, trying hard not to laugh. "Here's what we're going to do. I have a tow strap."

White smiles showed through the sweat-streaked mud on their faces.

"If you want me to take you to Meridian, you will stand together right here without moving while I get the Bronco and back up to your car. I will throw out the tow strap and one of you will pick it up. Then the other two will get into your car while one of you attaches the strap to the Bronco and then to your car. Then that person will get into your car. OK?"

"Yes, sure, fine, great," came the answers.

"You will attach the tow straps very carefully."

"You bet. Will do. No problem."

"I will stand over here and watch you attach the strap," I said. "You will stay in your car while I get back into the Bronco. Then I will tow you to Meridian and leave you at the first place where I can get us safely off the road."

"But—" one of them, Carl, protested. "Do you know where we are?"

"Sure," I said. "Since I'm towing you, it's going to be slow. I'm hoping I can get you back by eight. And it's going to be bumpy. Very bumpy."

All three of them nodded.

"If any of you do anything to scare me, I'll take off in the Bronco and batter you until you are meat in a bashed up tin can and your axles are left in the dust."

"We will do exactly what you say, won't we, guys?" said the tallest guy. "You're our angel. We'll do exactly what you say."

Two of them got into the car immediately. I backed up the Bronco and tossed the tow strap out into the winterfat. The first fellow picked it up and with the aid of my flashlight, attached the yellow tow strap to both vehicles. Quickly, he slid himself into the driver's seat of the muscle car, slammed the door and called out, "Good to go!"

"I'm leaving my window open," I told them. "Shout if something is wrong and you need me to stop. Otherwise, I'm going all the way to Meridian." I climbed into the Bronco and took off—very, very slowly.

The car, even with the added weight of the three men, popped right out of the dust blowout. I heard them cheering. We were on our way.

Towing the car through the desert was no picnic. As the winterfat, shadscale, and sage stands rolled by slowly, countless times I heard the undercarriage of the muscle car scraping dirt and shrubs—and shrieking as it skimmed over lava rocks.

The three guys sat tight.

Another badger glared at me from under a sagebrush, then retreated into a burrow. Short-eared owls, the phantoms of the winter desert, glided across the track more than once, but I saw no more mice, no more kangaroo rats.

The rising three-quarter moon cast a long, moving shadow to the west of the Bronco. I thought of the Snake River Canyon, only two or three miles away, gaping like a 500-foot-deep monster between black basalt cliffs, the river flowing winter-slow at the bottom, ice-rimmed now, the fish there sluggish, barely moving.

I cranked up the heat in the Bronco, but with a window open, I was cold and getting colder.

The sky was still black when I got to Swan Falls Road. I sped up to thirty miles an hour. The road was empty of vehicles. The only movement I saw was a big coyote flashing across fifty yards ahead, near the turnoff to Initial Point.

The town of Kuna slept soundly. We slowed to its speed limit and didn't see a single person, car, or dog. Back in the farmland, I turned north onto Meridian Road, and gradually increased my speed to forty miles an hour. Here we passed by a few cars, but the crush of the morning commute had not yet begun. *Talk about luck!* I thought.

When I arrived in Meridian at last, the sky was still dark. I pulled into a Quick Mart's empty parking lot, noticing a brightly lit phone booth nearby. Stiffly, I unfolded myself from the Bronco. The drive had gone on for hours.

The three fellows jumped out, and one began to undo the tow strap. Another one of the guys was already in the phone booth, calling someone.

"How can we thank you?" the tallest one said, handing me my tow strap. "I couldn't believe it when I saw your lights way out there in the middle of nowhere. Man, I thought we were toast. I thought my job was toast."

"I couldn't believe it when I saw the three of you," I told him, and we both laughed.

"Can I pay you?" he persisted.

"Tell the story to your children, when you have some," I said. Under the lone streetlamp, I could see his face for the first time, a handsome fellow under the dirt mask.

"I have to get to work, too," I said.

My office was less than ten minutes away. I'd have time to change into my work clothes there and tidy up a bit before eight o'clock. By pure luck, the skirt I had worn the day before was a reversible plaid-to-print, the top a simple black knit, so it wouldn't look as if I were wearing the same outfit two days in a row.

In the office bathroom, I changed and pulled my hair into a ponytail. My glasses were covered in so much dust that it was a miracle that I'd been able to see to drive.

Someone had come to work early and made coffee, another miracle. I poured my cup full and with a deep sigh, shoved the kangaroo rat boxes into the shadows under my desk, and sank gratefully into my soft chair.

The executive editor walked by. I might have known it; Bruce often arrived early and made coffee for the editorial group.

"You're here bright and early," he remarked as he passed.

I thought for a moment of what had happened during the night and how it could have turned out. "I may be early," I muttered, "but I'm not too bright."

From Rocky Wash to the Pentagon

I n the desert west of our house, there's a place we call Rocky Wash. Rocky Wash is just a lava-rock-floored dry streambed. It hasn't run water in the forty years I've known the place, but I imagine that back in the Pleistocene, it got plenty of liquid business, judging from the rounded shapes of the rocks there. Most of the year, Rocky Wash is a long, slight depression between stands of weeds and sage-winterfat mosaic, nothing distinguished about it.

But for two weeks a year, Rocky Wash blooms—locoweed and penstemon, scarlet globemallow and yellow mustard, purple asters and the beautiful plumes of Thurber's needlegrass, and coloring it all, the bright-yellow heads of sulfur buckwheat. In

Scott, taking photos for the Idaho Army National
Guard in Rocky Wash.

late spring, Rocky Wash is a river of gold. I can hardly imagine a place farther in every respect from Washington, D.C.

But a handful of years after I had been hired as a biologist by the Idaho Army National Guard, we learned of a competition. The Assistant Secretary of the Army for Installations, Susan Livingstone, was very concerned about preserving the ecology of military training areas. She would reward the three best programs with a trip to Washington, D.C., where we would speak about our programs in the Pentagon, meet the Commanding General of the Army—and the program that was voted the best would receive $50,000 to be spent on protecting or improving the biological health of that installation. The runners-up would each receive $30,000.

At that time, our program essentially had a staff of three: our colonel, Lt Colonel Sheehan, my supervisor Marjorie, and me. We were small in number, but we were one of the first environmental staffs in the entire Army system, important because our training area was not only one of the very best, but also because it was wholly contained within the newly designated Snake River Birds of Prey National Conservation Area.

My husband Scott took the materials we gave him and prepared a package to send to the Assistant Secretary's office. A package primitive by today's standards, it was sophisticated for the time: a notebook with photos, maps, graphs, and text. We mailed it off and crossed our fingers.

AND—we were chosen as one of the three installations to present at the Pentagon!

Four of us went, Lt Colonel Dick Sheehan, Marj, me—and our commanding general, the Adjutant General of Idaho, General Darrell Manning. We rambled around the Pentagon and got to go sightseeing during much of that week: the Washington Monument, Lincoln Memorial, the Vietnam Wall. Marj and I took ourselves to Arlington National Cemetery, and there I located the stone of one of my heroes, Admiral Richard Byrd.

Finally, it was time for the presentations. Fort Riley, Kansas, was one of the chosen. Their presentation was scary lovely, and they had twice the staff that we did. Marj and I shifted uncomfortably in the hard wooden chairs of the Pentagon's auditorium as we listened. But the winning presentation was from the Marine training area Camp LeJeune, an installation with an environmental staff of over forty. Their presentation was a professional-grade film, with music and all the bells and whistles. *Sigh*. Our presentation was just a slide show (pre-PowerPoint, you know), narrated live on the spot by Marj and me. But we were thrilled to be chosen as runners-up. Gosh, what could we do with $30,000 for our program? A GPS system? Native grass seeds? A microscope? All these things came to pass. After a picture-taking session with General Sullivan in the office of the Army's top general, we floated out of the Pentagon that afternoon on cloud nine. General Manning took us all to dinner away from the city at a lovely place overlooking a river.

Then something happened that is burned into my memory. As we pushed back our chairs after dinner and prepared to leave, General Manning said, "I'm going to take you somewhere now." He wouldn't say where. Twilight deepened into darkness as the general drove us back into the city. We climbed out of the rental car and saw that we were in West Potomac Park, near the Thomas Jefferson Memorial.

"Jefferson has always been my hero," General Manning said quietly. "We're going in."

I would never have dared to be out in this unfamiliar, urban place alone at night, but standing between a former Air Force pilot with over a million hours of flight time (General Manning) and a former helicopter pilot of the Vietnam Era (Lt Colonel Sheehan), I could not have felt more secure.

The inside of the Memorial was half in spotlight, half in shadow.

General Manning guided us from exhibit to exhibit, and we discovered that our general not only revered Jefferson but had memorized many of his writings. And so came the moments I will never forget: standing in the dark beside the starkly lit statue of Jefferson and hearing General Manning recite, "We hold these truths to be self-evident, that all men are created equal . . ."

The following year, in the late spring, I received a call from General Manning. I picked him up in our government Jeep and drove him down the dusty roads to Rocky Wash. He had remembered the photos of the flowers in our presentation at the Pentagon and wanted to see Rocky Wash for himself.

Every spring after that, until he retired, I would take General Manning to Rocky Wash at the peak time for flowers, and he would spend half an hour walking through the stony wash and asking me their names and life histories. And as many times as I watched the general bending over the delicate faces of the flowers, I thought of Jefferson and the Louisiana Purchase, and of how this harsh, vast land that I love came to be ours.

The Blue Hour in Hidden Valley

The high desert of Idaho is tan and gray. Much of the natural shrubland has been burned, and these parts of the desert are tan for ten months of the year. It's the color of invasive, non-native annual weeds, weeds that live only a few weeks, set seed, and die in place, fluffy and flammable, ready to catch fire from the next available spark, whether from a lightning strike or a carelessly-tossed cigarette.

I like to see gray vistas in my desert. Gray means that the native shrubs are still present, sheltering wildflowers, mosses, grasses, and wildlife, and are less likely to burn than the introduced weeds.

My favorite shrub is winterfat, a low-growing, long-lived woody fellow of pale gray, covered with fuzzy white hairs and looking like a cushion of spun silver in certain lights. Winterfat produces pale pink, fluffy flowers in midsummer. I carried a bouquet of blooming winterfat on my wedding day. Lest you think of winterfat as something faery-like, I will add that another name for winterfat is "French fries of the desert." Winterfat packs many calories into its narrow leaves and stems, even during midwinter when desert creatures are having a hard time getting enough food to keep themselves alive. Antelope eat it, deer eat it, rabbits and hares eat it, chipmunks and ground squirrels also eat it. Even grasshoppers eat winterfat. An individual winterfat plant can live longer than a hundred years if it doesn't burn.

In the military training area where I was working, there were still thousands of acres of winterfat in the late 1980s, its tight roots holding the soil in place against the relentless winds. The interspaces between the shrubs were filled with native grasses, mosses, and wildflowers. In a few select places where the soils were just right, a mosaic of sagebrush and winterfat grew, a sight most beautiful to a high-desert biologist.

One of these special places occurred in the Artillery Impact Area, a space of about 45,000 acres with a dirt road circling it. Firing ranges had been placed around the perimeter in the 1950s, and the tanks, Bradley fighting vehicles, and other weapons, (TOW missiles, rifles, the guns of Paladins, plus grenades and such) are fired there to this day. Aiming at targets within the perimeter is an important part of Guard training, while small fire trucks wait nearby to put out any fires that may occur. The Guard has been using the Impact Area since 1952.

In 1989, the very best stand of winterfat-sage mosaic in the region occurred deep in the Impact Area, and I was determined to explore it and document its condition. It took me a while to convince command that I and my field crews could work in the Impact Area without getting ourselves killed by unexploded ordnance, but we signed up for some training and at last permission was granted.

In the summer of 1989, we began marking certain randomly selected spots with very small steel stakes so we could locate them every year, and we recorded which species occurred there, how abundant they were, how tall they were, and the condition of the ground itself—whether it was tracked by foot, vehicle, wildlife, or livestock; if the spot was bare ground, lichen; moss, grass, shrub, weed, herb, dead litter, and so on. Livestock watering tanks were not allowed inside the Impact Area then, and we found wonderful communities of native plants, almost pristine. We loved working in the Impact Area, and treasured the days when the firing ranges would go cold so we could work inside.

The special winterfat-sage mosaic area we named Hidden Valley, a shallow swale; you'd never know it was there until you were almost upon it or saw it from the air.

By the time we were able to set our permanent monitoring plots in Hidden Valley, it was late summer; that year's wildflowers had bloomed and gone. There was scarcely a weed to be seen. Noticing the many dried flower stalks, we vowed to come back in the spring, but that was not to happen.

Four years later, we had rain, lots of rain, in the late spring. It was a great wildflower year. My senior technician, Jay, and I were keen to get into the Impact Area and head to Hidden Valley with our cameras. Most of the soldiers saw the training area when it was hot and dry, without a green blade of grass or green leaf of anything, and certainly without flowers. Scott, who had by this time been hired by the Guard, created the Guard's safety and environmental protection materials and videos. He was planning a series of posters, and he wanted some spring photographs that would show the soldiers why the plants were worth protecting.

Jay and I walked over to the Range building to check the Impact Area firing schedule for early June. There were no cold days! We were devastated, because a wet spring comes to this desert only once every ten to twenty years. After some arguing, pleading, and promising, the officer in charge of Range Control, Colonel Crew, agreed to let us make a quick trip into Hidden Valley on a certain day the following week. But there was a catch—we would have to be out of the Impact Area before 8 a.m.

On the designated day, Jay and I arrived at Gowen Field Base in Boise at 5:30 a.m. We took off in our government Jeep Cherokee, armed with cameras, film, lunches, and hot tea in our thermoses. It would take us well over an hour to get to Hidden Valley. Sunrise would be around 6 a.m., but the sky was cloudy. Would we have good light? And would there be many flowers?

We traveled the diminishing series of dirt roads as fast as we dared. Some people that morning weren't so lucky; at the base of Christmas Mountain, we passed a minor accident. A Bradley fighting vehicle full of soldiers had driven into a ditch hidden in tall weeds, and soldiers were jumping out to assess the damage.

A coyote crossed the road opposite Bigfoot Butte, heading into the Impact Area from a different route. The federal coyote exterminators didn't have permission to enter the Impact Area. There were many coyote dens inside the area.

We turned onto the tiny dirt road that led into the southern heart of the Impact Area. The sun was up now, but behind clouds. This was our only chance to photograph Hidden Valley this year before the flowers dried up. We'd have to do the best we could.

Finally, we came over the slight rise at the western edge of Hidden Valley. The clouds were breaking up. Sunlight flooded the small valley, as if it knew we were arriving.

We stopped on the downhill slope for a long moment, astonished. The sagebrush-winterfat mosaic looked like a pool of water.

Hidden Valley was *blue*!

The indigo spires of larkspur were everywhere, engulfing the sagebrush and winterfat in a spiky carpet of pure color. In all my years in the desert, I had never seen the like.

Larkspur is poisonous to livestock. In the 1930s, the Civilian Conservation Corps built roads, canals, bridges, and other works as the Great Depression ate up many jobs. They planted trees in burned forests, built docks on lakeshores, and did many other good things. But one of the things the CCC did in Idaho was pull up the larkspur plants. The CCC did a surprisingly thorough job of it. However, they missed a few spots, and Jay and I had just discovered that Hidden Valley was one of them.

The clock was ticking, and the clouds were closing in again. Quickly, we shot several rolls of film. We would have spent the entire day taking pictures if we had been allowed to stay.

Finally, when we knew it would be dangerous to stay a moment longer, we drove the Cherokee west, then south out of the Impact Area. As we passed Bigfoot Butte on the way to the northern part of the training area where we would spend the rest of the day working, we heard the *BOOM! BOOM!* of an M1A1 Abrams battle tank's main gun. Firing had begun.

We had the film developed, and Scott created a poster from one of our photos. It soon appeared in offices all over Gowen Field Base, including my own.

Later, it was decided by higher powers in another agency, that livestock watering tanks would be allowed in the Impact Area. Where once cattle and sheep had ventured only a few hundred yards inside, now they grazed the entire acreage. Our plots recorded the changes. Wildflowers and native grasses declined and, in many places, blinked out, year by year. Highly flammable weeds invaded, year by year. Fires became more frequent and more extensive. With heavy hearts, we documented it all.

And one summer, in the hundred-degree heat of summer, Hidden Valley, now choked with weeds, burned.

Today there are only a few remnant winterfat and sage plants in Hidden Valley, and scattered clumps of the hardiest of the native grasses. The days of blue and silver are gone. Hidden Valley is weeds. Weeds that catch fire easily. Weeds that will burn again and again.

But on one glorious morning, we believed that in a wet spring, the flowers would bloom there forever.

Action on the Lone Star

Sometime in the '60s, a guy my Dad knew called him. "Stew, I want to talk to you about a claim," the man said. Leo Rice was his name. Dad was always ready to talk to someone about a claim, so not long afterward, they met.

The claim was a galena mining claim (galena is an ore of silver, lead, and zinc) high in the Boulder Mountains. Leo was selling it, and he wanted one hundred and sixty dollars. "It's called the Lone Star Lode," Leo said. "The old tunnel is gone," he told Dad. "I think there's been a rockslide that covered it up. I bought the claim from two men who said they worked it during the Great Depression and on foot, backpacked enough high-grade material out of there to support their families. So, it's got promise."

Dad was most interested. And he was even more interested when he found that the claim covered almost 18 acres and was patented. He bought the claim.

These days, it takes a lot of effort and funding to patent a mining claim—construction work, assays, proof that working the claim will make a profit, environmental-effects assessments, and many more details. Claims are rarely patented these days. Patenting has become extremely expensive and time-consuming—impossibly expensive for the ordinary person today. But in the old days, patenting was easier. When a mining claim is patented, you own that land. You can do anything you'd like with it: mine it, build a motel on it, build yourself a house, dig a huge hole and fill it back up, plant it, fence it—just about anything. In fact, ownership of patented ground conveys more ownership rights than ownership of an ordinary home lot. Typically, homeowners own the house and land, but don't own the mineral rights as well.

The Lone Star Lode property is on a steep slope near one of the creeks and not far from the old Galena store at the base of Galena Summit in the Boulder Mountains. A rocky dirt road goes up that creek a mile or two, but soon peters out in the pines and firs.

The Lone Star Lode's hidden tunnel lies somewhere high in Idaho's Boulder Mountains.

Armed with the claim map, Dad walked up past the end of the road several times trying to find the remains of the old tunnel, but was never sure that he had found it. "There's a steep talus slope there, Danny," he told me. "It looks like it's slid a time or two, and I think that's where the tunnel was. It's now under probably ten feet of loose rock." Dad has always liked owning claims and for years, he kept the Lone Star, paying the annual property taxes of less than ten dollars a year. Knowing Dad, I think he had that dream all prospectors cherish: that he would find the old tunnel someday and be rich—or if not rich, at least satisfied.

In 1989, on my birthday, Dad signed the Lone Star Lode over to me. Scott and I kept putting off visiting the claim, but in the early '90s, after snowmelt was done and the roads were dry enough, Scott and I, armed with that same old map, decided to hunt for the tunnel. We drove our 4x4 a little way past the end of the actual road, and left it parked in a dense grove of pines.

The day was bright and hot. We were not certain we'd find

the tunnel, but in any case, the strong scent of pine and fir with the sun glinting off the tumbled rocks of the steep talus slope—those were treasure enough for us. And, besides, there were pikas.

I have many favorite animals, but pikas are in the top ten. Imagine a miniature bunny with teddy-bear ears instead of rabbit ears, and you have a pika. Pikas are not rodents, but are true rabbit relatives. They inhabit talus slopes near high mountain meadows, and besides being almost intolerably cute, they have excellent work habits.

All summer long, pikas harvest bundles of grass and herbs from the nearby meadows, spreading the bundles out to dry on flat rocks. If a sudden shower threatens, they will gather up the harvest and shove it under a rock; then when the rain is over, they will spread it out once more. When the harvested plants have dried to a crisp, the pikas take the bundles to their burrows under the rocks, storing food for the long winter to come.

One of the things I like best about pikas is the alarm call. When they see you, they will sit up on a rock and cry *"Geek! Geek!"* When you start climbing into the rocks, the pikas close to you will dive silently under the rocks, but the pikas above you will geek at you until you get too close to them.

To a rolling chorus of *geek*, Scott and I began to climb the long, steep talus slope on the Lone Star Lode property. We kept about twenty feet apart, looking for old timbers, pieces of metal, or anything that might indicate the presence of a tunnel under the rocks.

A shadow passed over us, and we heard a chorus of frantic geeks from all over the slope. Suddenly, something dark fell between us and lifted away with an indefinable *CRASH* of sound, something too speed-blurred for us to tell what it was.

We looked up and saw a golden eagle silhouetted against the sun. The eagle had dived onto a pika between us. The prey stoop of a golden eagle has been measured at 150 miles per hour, and that fellow was *moving*.

We never did find the old tunnel, but we still own the Lone Star Lode, an excellent pika sanctuary high in the Boulder Mountains.

Up Periscope

In the high desert of southwestern Idaho, it can be hot in midsummer. Infernally hot. Hundred-degree days are common, and sometimes temperatures climb to over a hundred and ten.

When I first began working for the Idaho Army National Guard, I and my field crew drove Jeep Cherokees, which held up admirably to long days on the rutted and rocky dirt roads, the heat, and the long, long days. But, during those first years of all-summer data collecting, those Cherokees didn't have air conditioning. We'd jump out of our Jeeps and spend an hour or so taking data at one of our permanent plots. We would locate the plot (in pre-GPS days it was not easy to find each tiny steel marker stake in more than 140,000 acres), roll out the hundred-meter tape, attach it at twenty-five meter intervals to the iron pins that were already in place, pull out the clipboard with its printed forms, carry the decimeter rod, and identify, measure, and record every plant that touched the tape. In later years we would use electronic data collectors, but the measuring and the heat were the same.

Oftentimes by lunch, we'd be as baked as a lasagna. If we were anywhere close to a high point, Jay and I would drive there to eat our lunch, resting for those few minutes in the hot shade inside the Cherokee. Sometimes, we would be lucky enough to catch a breeze at the top of a little butte. At other times, we would splash water onto ourselves and steam as we ate our packed lunches.

This particularly early July day was a scorcher, well over 110 degrees. By noon, we were barbequed. And the last plot we had managed to monitor was one at the foot of Bigfoot Butte.

Hoping for a breeze at the top, we headed up there for lunch.

The National Guard trains at Orchard Combat Training Center twelve months of the year, but in the early summer, the training is most intense, and the huge M1A1 Abrams battle tanks are on the move. An M1 tank is a tracked vehicle weighing over seventy tons, and on dirt roads, a convoy of tanks can turn the most compact soil into inches-deep dust, powdered finer than sifted cake flour.

The M1s had driven up Bigfoot Butte, and recently. Our white Cherokee churned clouds of pale dust as we fought up the steep road to the top of the butte. Even when we reached the top, the puffy dust filled the little road six inches deep.

Jay turned off the engine and we rolled down all the windows. If you keep the windows rolled up when you're driving through dust like that; you bake, but it's better than choking. And yes, we changed the air filters in our Cherokees often!

We grabbed our lunches from the cooler. I cracked open a cold Diet Coke and Jay uncapped a bottle of grape juice. *Ahhh.*

We saw a flicker of movement a few feet in front of the Cherokee.

A chipmunk sat on a rock on at the edge of the dust-filled road. He jerked his long tail this way and that and sat on that rock for some time.

Then he seemed to take a deep breath, and in the next instant, he plunged into the road. The dust swallowed him entirely.

A moment later, the terminal inch of his tail appeared above the dust. The tip of the tail plowed steadily across the road, leaving a long v-ripple in its wake.

Then the chipmunk bounced out and sat on a flat rock on the far side of the road. He shook clouds of dust from his striped fur and coughed for some time. As we ate our sandwiches, we watched him lick his paws and groom his dusty fur until it gleamed. With a final twitch of his tail, the chipmunk ran off into the shadscale.

Reluctantly, we fired up the Cherokee and drove back down into the simmering desert.

In Spades

It's easy to find advocates for wolves and cougars, elk, bears, and wolverines. But what about the small and secretive animals, the unusual and unattractive creatures of the world? In the world's ecosystems, there are wheels within wheels, and even the smallest wheels help the world go round. I've always been an advocate for the unnoticed and uncared-for, and one of these is the Great Basin spadefoot toad, *Spea intermontana*.

Great Basin spadefoots used to be more common in Idaho. MUCH more common.

My grandfather Orla believed that individual spadefoots were thousands of years old. He told me that one summer day in the early 1920s he was digging postholes for a pasture fence in the

My grandfather, Orla Hicks. Photo by Vicki Smith.

desert not far from Glenns Ferry. He had dug down two feet, chipping away at what he called "solid rock," when the rock cracked and out jumped a toad! He said that the toad croaked once and an instant later, died. He thought that the toad must have been there when the rock was forming, and therefore was older than stone itself. Of course, this is not true. Spadefoots can worm their way to amazing depths by following cracks and fissures. But when I was a child, I took this story to heart and felt that spadefoots were enchanted, almost immortal. I looked everywhere for them, but near our mountain town at 6,000 feet, I never found one.

Spadefoots burrow. On each of their hind feet is a "spade," a dark comma of hard material made of the same stuff as fingernails and hooves. You might picture spadefoots digging with their front feet, but that's not how they do it. A spadefoot will sit on a patch of dirt and stare innocently at you, but all the while, its back legs will be working. After a minute or two, you'll realize that the toad has dug himself almost out of sight. The last of the spadefoot that you'll see as one disappears into the dirt are those big golden eyes.

But why do spadefoots hop and dig? And how can a toad, an amphibian with moist skin and tadpoles that swim, make it in the Great Basin Desert?

This is how they live: typically, thunderstorms will move through the Great Basin in the spring, when the spadefoots are underground. When the spadefoots hear the thunder and feel spring warmth relaxing the bitter cold of the soil, they dig out and hop along the ground, seeking water.

The desert will green up, and for a few short weeks, intermittent streams will thread through the sage. Snowmelt puddles will form in hollows. Playa lakes will fill and support a multitude of tiny lives that can be food for tadpoles—daphnia, fairy shrimp, fly larvae, ostracods, copepods, and more. Male spadefoots will go in search of water and when they find it, they begin calling every night: *Rrraaaack! Rrraaaack! Rrraaaacckkk!*

Females and other males will come to the sound of the calls. They will mate, females will lay eggs, and then all will be off to fill themselves with insects and worms so that when they dig down once more, they'll have enough calories to survive until the next spring.

Sometime in June, everything dries up. The streams vanish and puddles disappear. Playa lakes become hardpan flats, crazed with cracks. But spadefoots dig deep into the soil. Each toad will line a small hibernation closet with mucus and sleep its long, long sleep.

Luckily for spadefoots, their tadpoles grow and develop with lightning speed. Eggs will hatch in two or three days, and full development takes only a month! If their home puddle is drying up, the tadpoles will eat one another in an attempt to develop fast enough to beat the drying of the remaining puddles.

Extensive, long-term grazing by livestock has changed the Great Basin Desert. Trampling alters the structure of soils and compacts it, so oxygen may not reach the level of the hibernating spadefoots in enough concentration to keep them alive. Much of this desert now has no spadefoots.

But my field crews and I found some.

Our work in the desert brought us out on muddy dirt roads in the early spring, places where there were snowmelt pools, temporary streams, and playas filled with water. In a few of those places, we found spadefoot tadpoles.

It was great fun raising the tadpoles in our lab. Tadpoles of some species of frogs and toads are picky eaters, and it's very difficult to get them to thrive. Spadefoot tadpoles will eat *anything*—fish food, worms, beetles, lettuce, bread, lunch-meat . . . (Please feed them fish food and worms if you catch any!)

Great Basin weather is extremely variable. During dry years, we wouldn't find a single tadpole. During wet years, we would find a few and grow some out in our lab, then release the tiny creatures back into the desert as soon as they had turned into toads.

The year after I retired, Scott and I were taking a drive in the desert and found a drying puddle in the middle of a small dirt road. In the muddy water were about forty spadefoot tadpoles helplessly wriggling, their backs already out of water, drying in the sun. I knew that in a day or two, the water would be gone, and this batch of tadpoles would die, all of them. So, we took them with us.

We have two 120-gallon Rubbermaid stock tanks at home. We filled one of these and released the tadpoles into the water. They grew fast, thriving on goldfish food, algae, and any insects they could grab. Every day or two, I'd walk out to the spot where we had placed the tank to see how they were developing, planning to drive them back to their home location when they had turned into toads.

But life got busy, and for a few days I forgot about visiting the tadpoles. When I remembered, the tank was empty. The spadefoots had gone, every one of them. "Stupid," I told myself. "Stupid."

Many months later, I had forgotten the spadefoots and their unknown fate.

One spring night I was coming in from scrubbing and refilling the two water buckets in the yard for the dogs after the dogs had come into the house, a task I do every night. I stopped on the front patio for a last look at the moon and stars, and I heard a strangely familiar sound.

Rrrraaaackk! Rrraaaack! Rraaaaack!

Intermountain spadefoot toad, a descendant
of the rescued tadpoles.

A Word from
Israel Russell

The year was 1901, and geologist Israel Russell rode up a certain sagebrush covered hill on his Appaloosa gelding, leading a dark bay loaded with the cumbersome photographic equipment of the day—a huge, heavy camera and large glass plates wrapped in black cloth. Russell was intent on documenting the salient features of Idaho geology for the first time. It was his job; Russell worked for the U.S. Geological Survey. His work resulted in a book with black and white photographs from that day's trek and many others: *The Geology of Idaho*.

On this day Russell was working south of the Oregon Trail and west of the small town of Mountain Home, in the sagebrush desert country. He had heard locals tell of two steep-walled depressions in black basalt, rounded and deep and sheer, like craters on the moon. A look at these would be well worth his time.

Russell toiled up a long, gradual slope with his two horses. Just when he was about to turn aside in disappointment, he saw craters opening nearly beneath his feet, Little Crater and Big Crater, two impossible voids punched somehow into the gentle sage hills, flat-bottomed 300 feet below, one crater oval, the other crater perfectly round. Russell got out his camera, set it up, and took a photograph.

I have a copy of that photograph on my desk, leaning against my scanner. Russell's photo shows sagebrush on all sides as far as his camera could see, Cinder Cone Butte a vague but recognizable bump on the far and strangely ghostly horizon. The foreground reveals the edges of the two deep craters, shadowless, open to the sky. Sagebrush crowds the rims of the craters, and like a cloak of gray velvet, covers everything in the landscape except the black basalt rock of the crater cliffs themselves.

Much of the sage is now gone.

With a mind to craft plans for ecosystem restoration, I began collecting photos of the western Snake River Plain as it used to be, and at once came upon Russell's work. He and other pioneer photographers had taken many images relevant to my project, but every time I looked through them, I found that I was lingering over Israel Russell's image of Crater Rings.

Crater Rings, Elmore County, ID, looking west toward Cinder Cone Butte; photo by Israel Russell before 1901.

Suddenly I had an idea: I would revisit the sites of all those old photos, re-photograph them, and document the changes in the plants! First on my list, of course, was Crater Rings.

I had been to Craters before, so I knew the way. Traveling east on the freeway from Boise, I turned south onto Simco Road, east again on Cinder Butte Road, through the farm flats, and up into the dry hills—paved highway to gravel road, gravel road to dirt road, dirt road to dirt track. Then I looked for a tiny, very rocky thread of road to the southeast, where there is a loop gate in the barbed-wire fence. I went through the loop gate and up the hill to another loop gate.

I always had to smile when I went through this particular gate on the hill. The loops for holding the gate shut were large, not too tight, and made of heavy, smooth wire. Two wooden signs nailed to fenceposts there bade travelers to please close the gate to keep cattle where they belonged. This was one of the gates in Domingo Aguirre's federal-land grazing allotment, and Domingo was smart.

I first met Domingo at this very gate. He had been riding a stunning flaxen red sorrel gelding, very flashy and with quality rigging. Domingo himself was no less colorful in jeans, a purple satin shirt with white scrolling, and a rakish neckcloth of turquoise blue. "My gates are good gates," he told me, slipping from the saddle and leaning back against his horse's shoulder in the bright spring sunshine. "If someone, maybe a woman like you, comes to one of these tight, macho barbed-wire gates and gets it open, do you think she would want to shut it again? Do you think she would even be able to shut it? No, no, I say. But my gates I make easy to open, and I ask politely on the signs for them to be closed. And people always close my gates, and my cattle never get out. You see?" I saw. With that, Domingo had loaded his beautiful sorrel into a waiting horse trailer, climbed into the attached pickup, and disappeared into his own dust cloud.

In 1981, I saw no one on the hill of Crater Rings. Spring was long past, and the non-native plants were pale and sere. Not even cattle braved the drying wind. I opened Domingo's good gate, closed it dutifully, and drove carefully across a stretch of angular lava boulders to the top of the hill.

Stopping the truck on a level space of black rock, I took out my camera and a copy of Israel Russell's old photograph of Crater Rings. Now to frame, as accurately as I could manage, the same view that Russell had seen through his lens. Of course, my camera had a different format from the one he had used, so the framing would necessarily be slightly different. Nevertheless, I wanted my photograph to show, as closely as possible, the same view that Russell had documented.

I walked back and forth across the top of the Crater Rings hill, stumbling over rocks while peering through my viewfinder. "Now, look," I said to myself, "his photo shows the southern edges of both craters, and Cinder Cone Butte is directly behind the southernmost point of Little Crater's edge, dead-center in the middle of the background. I must find where he stood!" I found myself walking west, slightly down from the crown of the hill.

In my path loomed a small outcrop of broken lava, almost as tall as I was, about the same size as a speaker's podium. As if on automatic pilot, I walked toward the outcrop, camera extended.

The crumbling top of the outcrop contained a distinct niche, floored by a natural platform of stone, small but level, at the height of my shoulder. With a kind of electric awareness, I set my camera on this little place and looked through the viewfinder.

Here was Israel Russell's photograph!

I felt a connection with the long-dead geologist click into place as neatly as a camera shutter snapping into its groove. Russell had stood here exactly. He had placed his own camera on this spot, in this rock, and had pressed the shutter just as I was doing at this moment.

Eighty years had passed since Israel Russell stood here, but I felt him standing beside me.

His sea of gray sage had nearly gone.

In its place was a sweeping carpet of alien, highly flammable plants the color of dead grass, plants introduced from the Mediterranean, the Gobi Desert, and Europe, aided in their takeover by decades of livestock grazing and fires.

Cinder Cone Butte was now naked of shrubs and crowned by two communications towers that stood crisply outlined and blunt on the distant horizon, nothing ghostly about it in the latter part of the 20th Century. My photograph shows it well, along with the precise and harsh line of the horizon and the pale plants that do not belong.

Twenty years more fled by, and the year 2001 was upon me, with my friend and senior technician Jay Weaver, standing at Russell's Rock taking yet another photograph in the series of then and now—exactly a hundred years after Russell.

The sage had gone altogether, and the yellow alien grasses were bent nearly flat to the ground by the cold wind of early winter. It was midday. The wind bit at our hands and faces as I placed yet another camera in the little stone niche and took another photograph of the edges of the craters and of Cinder Cone Butte.

The shutter clicked and we were done. We lowered our heads into the wind as we made our way back to our pickup, which we had left parked near the edge of Big Crater. I found myself wondering what Russell would make of this landscape of the 21st century. I wondered what he would say. But Russell had been a geologist, not a biologist. The craters would be the same.

Just ahead of me, Jay had stopped in his tracks. I almost ran into his back, lost in thought as I was. And then I heard it: a strange murmuring and clicking, a vast swishing sound I had never heard before, and have never heard since. Both of us cast about in a circle looking for the cause.

We were alone at the top of the long hill. We could see for miles, and there was nothing in our world but sky, rock, alien weeds, and our solitary pickup. The sound grew louder, became almost deafening.

Up from the crater, with incredible velocity, spun a black whirlwind.

"Ravens!" one of us cried. They rose from the crater suddenly, wings beating and beaks clicking, a black cloud made from hundreds and hundreds of Ravens.

Jay and I had worked together in that desert for many years and never had we seen the like. The spinning vortex of ravens rose a hundred feet into the air above Big Crater. We stood on the brink as if cast in stone, astonished and wordless.

The black tower of wings spiraled higher and higher into the gray sky with a sound like the breathing of a monster. The ravens whirled madly for a minute or two more. Then one raven uttered a single croak. The maelstrom of birds broke, wheeled, and scattered across the sky.

We turned our backs on the landscape of alien weeds and shut ourselves into the pickup. Carefully, I laid Russell's photograph of the sea of sagebrush on the dashboard and reached for my thermos of hot tea.

A Day in the Life:
Letter to Marie, 17 June 2003

Dear Marie,

This weekend Chris Osborn from California drove up to meet us and look at the Shelties. She is looking for a very bright male for agility competition who is built well and is pretty enough for conformation. Chris arrived Saturday night. She liked Raleigh very much, and said he was the best built dog she had seen in some time (he really does have awesome running gear).

Then the next day, Sunday, one of the soldier liaisons, I'll call him Max, called saying that I had to go into the Artillery Impact Area immediately to rescue a nest of baby ravens, since the old, wrecked vehicle they were in was going to be used as a target as part of a tank-firing exercise. So, Chris and Scott and I jumped

into the work Jeep and took off. We went out there into all the heat and dust, as it's the middle of the Guard's major summer two weeks of military training. Max said that he'd been told the baby ravens had been abandoned by their parents and were at death's door, extremely weak and dying.

Well, we got there and found four baby ravens in the shade, up inside the gutted remains of an old Army personnel carrier. They were just fine—fat, strong, and healthy, and their parents were there, flapping above us and anxious. But Max said that the carrier was going to be shelled right away.

So, I gave the unsightly babies (baby ravens will never win any beauty prizes!!) each a slug of Pepsi and put them into a plastic washtub, and we got out of there. I later found that the Range Control soldiers were perfectly willing to shift targets to something a couple of miles away so that the nest would not be disturbed. So, I will have words with someone (not sure which person) tomorrow since we should not have taken them at all. I would have loved to have had our Sunday off, sigh, and I have no idea what Chris thought! We got back five hours later, established the ravens in our old pheasant cage and fed them more Pepsi and some hot dog bits, and I then put all the dogs up on the table for Chris to examine and ran them around so she could see their gaits. Then she departed, and I took a nap.

The next day we took the nestlings to the Animals in Distress Rescue, and the rescue people loved them. Since baby ravens do well in captivity, I am sure they will be fine, and hopefully we can be in on the release when they are grown.

This morning, Monday, when I got up, I had a migraine, a bad one. I called in sick to the office since I don't dare drive during one of these episodes. But dog chores have to be done every day. I let Ramona into the yard with Coalene, Rowena, and Raleigh to potty, and told myself to bring Ramona right in because her temperature was down, and I figured she would have her puppies today. Then I got tied up on the phone. Twenty minutes later I staggered out to bring her in—and she'd had two puppies! She and Coalene were surrounding them, both licking them and curled up like both halves of an 'O' with the puppies in between. Rowena and Raleigh were lying down nearby looking on intently.

The puppies were fine, all cleaned up. So, I grabbed them and Ramona, and put them into the prepared box in the kitchen, and she had the final one. She just squirted them out in no time. All five were born in the space of half an hour. We have a blue male, a blue female, two tricolor males, and a tricolor female, sired by International Champion Cherden Thunder Run. They aren't very uniform (what you would expect with a line cross) but are quite pretty. The tricolor girl is fat and the largest, the blue girl in the middle, and the tri male is the smallest. The blues are dark blues with traditional merling patterns and not much black at all. I am quite pleased with them, and Ramona is being a good mom.

Ramona's litter of Sheltie puppies.
Two of them became AKC champions of record.

Then on top of that, at the end of the day Lacey called from work and said that one of my field crews, I'll call them Ron and Larry, had not come in and she could not get them on the radio. So, I drove in to work at 6 p.m., planning to get into my Jeep and search for them. By the time I arrived at the office, Lacey told me that Ron and Larry were being brought in by soldiers, because they had wrecked their government Jeep. *Arrrgggh*. Thank God they were OK. They had driven a very steep dirt track down into the Snake River Canyon, had realized it was too steep, turned around and tried to get back up to the rim, failed, and rolled the Jeep. They had their seat belts on, yay!! That happened in the "hole" where cell phones don't work. It was too dangerous for them to get to the Jeep's radio, so they walked five miles to where some soldiers were training and got a ride in. Since the canyon is about 600 feet deep at that point, they could have been crushed; I shudder to think of it. Tomorrow the soldiers will retrieve the Jeep, which Ron says is totaled. *Arrrrgh* again! Tomorrow we will have stacks of paperwork, and the Range Colonel will send a helicopter to lift the Jeep out of the canyon. So glad my people are OK. We always have one of us stay in the office until every last person is in from the desert, so bless Lacey's heart for staying. I will pick up some flowers for her tomorrow.

Well, must go feed the bow-wows before I crash. Hope everything is fine with you!

Love,
Dana

Hiss and Growl

One fine spring day, Jay and I were driving south on Standifer Road, a halfway decent gravel road in the Orchard Combat Training Area.

We had a bunch of rare plant monitoring to do for a study documenting, and for the first time, the life history of slickspot peppergrass, *Lepidium papilliferum*. The name of the plant is misleading since it's not a grass. We called it LEPA. LEPA has clusters of tiny white flowers and looks rather like sweet alyssum, but LEPA belongs to the mustard family and is NOT sweet. This work was to result in our paper "A life history of the Snake River Plains endemic *Lepidium papilliferum*," co-authored with Dr. Susan Meyer and published in *Western North American Naturalist* in 2004.

On the way to the LEPA study area, we noticed a brown lump at the side of the road. I thought at first that it was a dead ground squirrel. Then it moved, and Jay pulled our government Jeep Cherokee to a stop.

The brown lump was a baby badger, so young that it was still cutting its milk teeth. Jay lifted the baby from the ground and turned it over: a female. Something was wrong with her—she had a deep wound on her back. We found a box in the back of the Cherokee and put her in it. "We had better search to see if there are any more cubs," Jay said, so we did just that.

After a few minutes of searching the area, Jay found the body of an adult female badger, one that had been nursing. She had been shot. A few minutes later, we discovered a second baby badger. He tried to get away, but we cut him off as he scurried toward a burrow. This fellow was uninjured and hissed and growled mightily. We put him into the box with his sister and headed for town.

Once we got to my office, we gave the babies some water and called the "badger lady," Mady. Mady and her sister Toni are the founders of Animals in Distress, an organization that cares for injured and orphaned wild animals. They are both expert and tireless, and are armed with compassion, knowledge, and the proper permits for housing and rehabilitating wild creatures. I had dealt with them before and called Mady at once.

Jay and I drove to her house as soon as I ended the call.

"That female has been attacked by something," Mady said after a quick look at the smaller furry baby. "As soon as I get some formula down both of them, I'm taking her to the vet."

Knowing that the baby badgers were in good hands, we climbed back into the Jeep and were soon heading south on Standifer Road toward our LEPA plots once more.

Both cubs survived. A few weeks later, we visited Mady's place and saw the half-grown siblings rooting and sniffing in a large dirt-floored pen. Their coats were sleek and shining, and they growled at us. Continuously.

Autumn rolled around.

Mady called me in early September. "It's time to release your badgers," she said. "I was wondering if we could let them go up your gulch. Is there water? Are there any prey animals? They won't be very good at finding water and at hunting for a while, and I thought they might be safer at your place than out in the desert."

"Sure," I said. "There's a spring just up the gulch from us, and there are plenty of ground squirrels and pocket gophers there."

A date was set, and on a warm Saturday afternoon, Mady and Toni drove to our house. Scott and I showed them where to park near the spring, about a hundred yards from our back door.

They had brought a chain-link dog run, with a chain-link top as well, plus a generous supply of fruit and frozen chicken breasts. "They will need to be fed for a while until they get used to the place," Mady told me as we helped her and Toni put the dog run together. With sturdy wire, we attached the top panel to the run so the badgers wouldn't be able to climb out. "After a few days you can open the door, but keep food and water in the run for at least a week, so they won't starve while they are learning to hunt," Mady advised.

Mady pulled a water pan and a food pan from the back of Toni's truck. "Next weekend we'll come back to check on them," she said. "Are you sure you won't mind feeding them twice a day and filling the water pan?"

"No problem," I said, adding dubiously, "Do you think they might dig out?"

"I don't think so," she said. "They've never tried that in their pen at home."

We filled the water pan, and Mady placed a cluster of grapes and two chicken breasts in the food dish.

When the chain-link run was ready, Mady and Toni returned to the truck. Two dog crates in the back held the badgers.

Mady pulled out the male. She held him with his back against her waist and his paws, with their gleaming three-inch claws, draped over her arm. He was a hefty fellow, perhaps thirty-five pounds. He glared at us and hissed. Growls rumbled from his throat as he snarled at Scott and me while Mady kissed the top of his head. Toni brought out the female. She was slightly smaller than the male but just as growly.

The growls reminded me of the woeful tale of a National Guard soldier from an out-of-state unit who found a badger near his unit's camp in the training area. This fellow had popped a cardboard box over the badger, then picked up the box and began to walk back to his camp to show the badger to his fellow soldiers. The badger promptly ripped the bottom out of the box and proceeded to tear the skin and some muscle from the soldier's thighs. The soldier took an unscheduled helicopter ride to the nearest hospital. One night another soldier, disdaining the portable outhouse at his bivouac, decided to poop in a convenient badger hole. The occupant was home and was quite miffed. That soldier, too, was life-flighted to Boise, with serious injuries to his bottom and private parts.

Badgers are not to be messed with! Working in the desert, I have picked up many a badger skull over the years. A badger is essentially a huge weasel, with a weasel's thick top skull, heavy jaws, and sturdy, cone-shaped teeth. You could probably stab a badger in the head with an ice pick and it would just irritate him.

Toni and Mady carried the growlers to the door of the chain-link run and with final kisses, set them down inside. We padlocked the door shut. With a wave, Mady and Toni drove away down the gulch.

Ignoring the food, the badgers began sniffing the perimeter of their new run. Scott and I carried their food supply to the house and stuffed it into our back-room fridge. I laid two of the frozen chicken breasts on the kitchen island to thaw for the badgers' dinner.

A couple of hours later, I slipped the chicken breasts into a plastic bag, and we walked up the gulch to the badger hotel.

They were gone.

The top panel of the chain-link run was lifted at one corner, just enough to allow an escape. They had climbed up the chain-link, had broken the heavy wire securing the top panel in place, and had gone, probably minutes after we left. We never saw them again.

I looked at the plastic bag I was carrying. "Well," I said to Scott. "I know what we're having for dinner."

Epilogue: What We Take

I have lived a small life.

I haven't climbed Everest or plumbed the depths of the Mariana Trench on my journey into wild habitats. But as you have discovered by now, I have dipped a toe into a few strange places and into a number of little lives. And I have learned much about the real world along the way.

Here is the last and most precious lesson the wild has taught me. Dad put it into words when I was five. Here's how it came to be:

***

I was a sophomore at the College of Idaho in the mid-'60s, home in Ketchum for the summer, working evenings at Sun Valley Resort's golf shed: girl golf cart mechanic at your service. I serviced the golf carts after sundown each evening, so the days were my own.

My boyfriend was hundreds of miles away that summer, working on a farm near Burns, Oregon. We communicated via letters. Eric knew he would be taking entomology in the autumn, and asked me to collect insects for him, a specimen or two of as many kinds of insects as possible. He would do the same. Since autumn frost comes early in Idaho, entomology students were encouraged to collect their specimens early, then learn to identify them during the course.

I knew just where I'd go once the weather warmed up: the North Fork.

The canyon of the North Fork of Wood River has always seemed strange. When I was a little girl, I felt that the North Fork was filled with ghosts. The upcanyon wind through the trees was a song, the gold-striped salamanders in the mouths of the tunnels, spirits.

Dad and Gramps bought the patented mining claim there when I was very small, a triangular acreage bounded by the river and a small creek on two sides and a mountain on the other. The property included several cabins, a dynamite shed, ponds, a marsh, hard-rock tunnels, and ore cars on rails that led far back into the unfathomable darkness of the mine.

There were two main tunnels into the mountain—the lower tunnel, padded with velvety green and red moss between its silver rails—the entrance seeped a runnel of water—and the upper tunnel, dry and rocky and rail-less.

Growing up, I loved North Fork. Dad or Mom would often drop me off to play there alone. The rule was this: Danny could go into the tunnels, but only as far as she could still see light from the entrances. I never went farther inside.

The only time I can remember Mom being very angry with Dad was the day he took me back into the mine.

It was the lower tunnel. He oiled the solid-metal wheels of the rusty ore car and lifted me inside. I was five years old and up for almost any adventure. Dad lit his carbide headlamp, gave the car a shove, and in we went.

Heavy timbers shored up the first part of the lower tunnel.

A few had miner's candle picks still embedded. In the first few yards of that journey, I remember the smell of wet rock and the sound of water dripping.

In mines, darkness is absolute. Shortly after we left the entrance, so was the silence.

The tunnel rose and fell. Dad put his shoulder to the heavy car as we went up the little rises and dragged his feet to slow its speed on the downslopes. We went past a gaping hole that seemed to swallow all the light from his headlamp. "That's a winze," he told me. "I've tossed rocks down it. It's really deep."

We went further. Cracks in the tunnel roof and walls had been common at first, but the farther we traveled, the fewer they became.

Here and there I began to see bright threads and seams.

Dad stopped to scratch at one of these with the point of his rock hammer. Silvery bits crumbled off the wall into his hand, tiny cubes shining in the light. "This is galena, Danny," he said, holding his lamp's beam on the shimmery heap, "galena ore. It contains lead and silver." He poured the crumbles into my hand, and I dropped them into my pocket. "That's right," he said. "You're a born miner. What we take from North Fork is worth keeping."

We went on, past branching tunnels and into a long, still darkness. Oddly, the air seemed to be getting warmer.

Eventually, Dad stopped the cart.

"How far have we gone?" I asked him then.

"I think we're about half a mile back into the mountain," he said. "We're turning back now. Ahead, there's some rockfall, and it's not safe there. But for just a minute, we're going to listen to the living rock. We're part of the mountain here, and I want you to remember this." He switched off his lamp.

Total darkness billowed around me. When I closed my eyes, odd shapes slid in and out of view across my eyelids. Then, I didn't understand how nerves work, so this was unsettling. I opened my eyes into blackness and held my breath. Silence.

Above us was suspended the unimaginable weight of millions of tons of rock and soil, of a mountain, of thousands of pines and firs—and that weight was prevented from crushing us by a simple tunnel blasted into the rock by miners with dynamite fifty years before. The blast grooves still remained in the rock; I had seen hundreds of such grooves, and indeed, a box of moldering dynamite had once been stored in the shed just outside the lower tunnel. Dad and Gramps had taken it away.

Dad turned on his lamp.

He moved to the far end of the car and began pushing it back the way we had come. Return journeys always seem shorter, don't they? Back we went, and I was relieved when I saw the grayness of the daylight at the end of the tunnel.

At the entrance, Mom was waiting.

"You took Danny back into the mine!" she shouted. "You didn't even tell me you were going!" There were tears in her eyes. "You were gone for hours, Stew, *hours*. I didn't know if you would ever come out. I was about to call Dad. I was giving you fifteen more minutes. I know you go back in there all the time, but not both of you, Stew! Not both of you at the same time!"

"I'm sorry, Bernice," he said. "I won't take Danny in there again." And he didn't.

***

Then I was in college.

Dad and Gramps had sold the mining claim a few years before, but every summer, I still came often to North Fork. Sometimes I would flyfish. Sometimes I would bring a book to read in the shade of one of the tall firs. Sometimes I'd bring a rock hammer and my rock bag. And sometimes I'd just walk and walk. North Fork is a cold canyon in a cold land, and its brief summers are precious.

And North Fork was and is strange. I found butterflies there that I never saw anywhere else, and I saw grasshoppers in the marsh that I never saw in any of the surrounding canyons and gulches.

So, this day I drove my gray 1960 Rambler, Coalie, off the main road, through Murdock Creek, and onto the mine property. I took my butterfly net from the back seat and set out to collect some insects for my boyfriend.

There was a small hill—more like a mound—at one edge of the marsh. Wild geranium grew there, and in the air above the flowers I could see butterflies fluttering. There was one in particular, rust-orange with silver markings, that I had to have.

Net in hand, I ran up the mound after the butterfly, and at the top, I collided with a man.

That man had a butterfly net, too. His net met my net in the air above our heads, we crashed together, and we both landed on our backsides, astonished.

"What are you doing?" we asked each other simultaneously. Then we both laughed. Another man joined us on the ground. Both men were young, scarcely older than my twenty years.

"We're from Utah State University," that man said. "We're graduate students there, studying mayflies."

"I saw a mayfly just now," the first man said. "I was going after it. We're getting kind of desperate," he explained. "We have been working along Wood River for over a week, and we were counting on collecting a bunch of mayflies for our dissertations, but we've come up empty. We've been back and forth all the way from Hailey to Galena Summit and haven't seen a single swarm. I was hoping we'd get a few here near this marsh."

I smiled. "I've got lots," I told them.

"Do you collect mayflies?"

"Nope," I told them, "but Coalie collected a bunch last night." As they followed me back to my car, I explained that the night before, I had been flyfishing farther up the main canyon, and on my way home in the twilight, I'd driven through a swarm of mayflies near the place where Baker Creek flows into Wood River.

The two graduate students spent an hour with forceps picking mayflies from Coalie's grille, while I went back after that butterfly. In mid-afternoon, we parted ways, and I headed home. They were heading upcanyon to the place where Baker Creek flows into Wood River.

That September, I gave my collected insects to my boyfriend. The rust-colored butterfly with the silver markings was a record, the only time that species had ever been collected in Idaho.

***

North Fork is a still a part of me, as are the mountains above Compostela, and Ravenshoe rainforest, Arctic National Wildlife Refuge, Bear Canyon, Fisher Creek, Big Pine Key, the Snake River Plain, and all the other places filled with the ghosts of old memories.

After so many years, Dad's words still come to me from the unforgotten blackness of the mine. *What we take—what we remember—is worth keeping.*

Acknowledgements

I would like to thank my husband Scott and other family, friends, colleagues—and my many students and field crew members—for their contributions to the little life (mine) described herein. Much appreciation goes to my editor, Kerstin Stokes, for her expertise and patience. Thanks to Eric Yensen for some corrections and timeline adjustments. The College of Idaho kindly provided the photograph of Dr. Lyle Stanford and students. Most especially, I'd like to thank my long-time friend and colleague Bill Clark for the use of his photos taken on the 1965 College of Idaho Mexico Expedition.

I wrote these stories as I remember. Any errors are mine alone.

To my readers: may you make many memories worth keeping.

About the Author

I grew up in Ketchum, Idaho, enchanted by wildflowers and the small lives of wild things, taught to fly fish by my father, a master. After college, I married and, among other things, worked as a scientific illustrator, a field research technician, a golf-cart mechanic, a high school teacher, a college instructor, co-leader of college field biology expeditions, and for years was the biologist for the Idaho Army National Guard, ending up as Natural and Cultural Resources Manager for the State of Idaho Military division. But titles don't tell much. Here are some of the things that made up my life: watching clownfish lay eggs on the Great Barrier Reef; documenting a species of intertidal ants in Mexico; planting thousands of acres of native plants after wildfires; discovering a predatory new species of fairy shrimp; watching arctic foxes and their kits foraging along the Arctic Sea; discovering the life cycle of a rare white flower; learning the mountains and hills, the deserts and canyons of the West. And writing.

Always writing.

Books by Dana Quinney:

Wildflower Girl (2019, Hidden Shelf Publishing House)

Explore the
Hidden Shelf

www.ingramcontent.com/pod-product-compliance
Lightning Source LLC
Chambersburg PA
CBHW061140120626
46546CB00005B/1867